REVISE EDEXCEL
A2 Mathematics
C3 C4

REVISION WORKBOOK

Series Consultant: Harry Smith

Author: Glyn Payne

Our revision resources are the smart choice for those revising for A2 Mathematics. This book covers Edexcel units Core Mathematics 3 and Core Mathematics 4.

- **Organise** your revision with the one-topic-per-page format

- **Speed up** your revision with helpful hints

- **Track** your revision progress with at-a-glance check boxes

- **Check** your understanding with guided questions

- **Develop** your exam technique with exam-style practice questions and full worked solutions.

Revision is more than just this Workbook!

Make sure that you have practised every topic covered in this book, with the accompanying A2 Mathematics Revision Guide. It gives you:

- More exam-style practice and a 1-to-1 page match with this Revision Workbook

- Worked examples to help build your confidence

- Hints to support your revision and practice

- 'You are the examiner!' sections to help you practise checking your work.

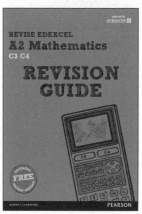

For the full range of Pearson revision titles across GCSE, AS/A Level and BTEC visit:
www.pearsonschools.co.uk/revise

ALWAYS LEARNING

PEARSON

Contents

A small bit of small print

Edexcel publishes Sample Assessment Material and the Specification on its website. This is the official content and this book should be used in conjunction with it. The questions in this book have been written to help you practise every topic in the book. Remember: the real exam questions may not look like this.

Mathematical Formulae and Statistical Tables

The Mathematical Formulae and Statistical Tables that you need for your exams is available from the Edexcel website.

Algebraic fractions

> **Guided**

1. Simplify fully $\dfrac{2x^2 + 7x - 30}{x^2 - 36}$ **(3)**

$$\frac{2x^2 + 7x - 30}{x^2 - 36} = \frac{(2x\ldots\ldots\ldots)(x\ldots\ldots\ldots)}{(x\ldots\ldots\ldots)(x\ldots\ldots\ldots)}$$

> Remember that $a^2 - b^2 = (a + b)(a - b)$.

$$= \frac{\ldots\ldots\ldots\ldots\ldots}{\ldots\ldots\ldots\ldots\ldots}$$

2. Simplify fully $\dfrac{6x^2 - 23x - 18}{2x^2 - 11x + 9}$ **(4)**

> Fully factorise the numerator and the denominator.

....

....

3. Simplify fully $\dfrac{2x^2 - 32}{5x^2 + 17x - 12}$ **(4)**

....

....

4. Express $\dfrac{3}{x(x - 1)} + \dfrac{2}{x}$ as a single fraction in its simplest form. **(2)**

> The LCM is $x(x - 1)$ because x is a factor of both denominators.

....

....

....

5. Express $\dfrac{3x^2 + 14x - 5}{x^2 - 25} - \dfrac{4x}{x(x - 5)}$ as a single fraction in its simplest form. **(5)**

> Factorise fully then look for a common denominator.

....

....

....

....

6 Express $\dfrac{2x - 3}{3x^2 - 7x - 6} + \dfrac{4}{3x + 2}$ as a single fraction in its simplest form. **(5)**

....

....

....

....

Algebraic division

Guided **1.** Use long division to divide $2x^4 - 7x^3 - 10x^2 + 24x + 10$ by $x^2 - 3$ **(4)**

> Remember to write $x^2 - 3$ as $x^2 + 0x - 3$.

$$
\begin{array}{r}
2x^2 \dotfill \\
x^2 + 0x - 3 \overline{)2x^4 - 7x^3 - 10x^2 + 24x + 10} \\
\underline{2x^4 + 0x^3 - 6x^2} \\
-7x^3 - 4x^2 + 24x + 10
\end{array}
$$

....

> Always line up the terms with the same power of x.

2. Given that

$$\frac{2x^4 - 5x^3 - x + 6}{x^2 - 2} \equiv ax^2 + bx + c + \frac{dx + e}{x^2 - 2}, \qquad x^2 \neq 2$$

find the values of the constants a, b, c, d and e.   **(4)**

> Multiply through by $(x^2 - 2)$, expand brackets and collect like terms, then compare coefficients, starting with the highest power of x.

....

....

....

....

....

....

....

....

3. Given that

$$\frac{5x^4 - 2x^3 - 3x^2 - 8}{x^2 - x - 2} \equiv ax^2 + bx + c + \frac{dx + e}{x^2 - x - 2} \qquad x \neq -1,\ x \neq 2$$

find the values of the constants a, b, c, d and e.   **(4)**

....

....

....

....

....

....

....

....**(4)**

Functions

Guided **1.** The functions f and g are defined by

$$f : x \mapsto x^2 - 2, \quad x \in \mathbb{R}, \ x^2 \neq 2 \qquad \text{and} \qquad g : x \mapsto 2 + \frac{3}{x}, \quad x \in \mathbb{R}, \ x \neq 0$$

(a) Show that the composite function gf is $gf : x \mapsto \dfrac{2x^2 - 1}{x^2 - 2}$ **(3)**

$$gf(x) = 2 + \frac{3}{x^2 - 2} = \frac{2(\ldots\ldots\ldots) + 3}{x^2 - 2} = \frac{\ldots\ldots\ldots\ldots\ldots}{\ldots\ldots\ldots\ldots\ldots}$$

(b) Find $gf\left(\dfrac{1}{2}\right)$. **(1)**

....

(c) Solve $gf(x) = 0$ **(2)**

....

2. The functions f and g are defined by

$$f : x \mapsto 1 - 5x, \quad x \in \mathbb{R} \qquad \text{and} \qquad g : x \mapsto \frac{2x}{x + 3}, \quad x \in \mathbb{R}, \ x \neq -3$$

(a) Show that the composite function gf is $gf : x \mapsto \dfrac{2 - 10x}{4 - 5x}$ **(2)**

....

....

(b) Find $gf(-4)$. **(1)**

....

(c) Work out the composite function gg. **(4)**

....

....

....

....

(d) Find $gg(-1)$. **(1)**

....

(e) Solve the equation $gf(x) = f(x)$ **(4)**

....

....

....

....

Graphs and range

> **Guided**

1. The function f has domain $-5 < x < 6$.
Here is a sketch of the graph of $y = f(x)$.

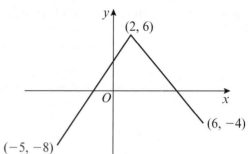

(a) Write down the range of f. **(1)**

Range of f is $-8 < f(x) < $ ....

(b) Find ff(6). **(2)**

ff(6) = f(−4) = ...

> Use the gradient to work out f(−4).

2. Find the range of the function defined by

(a) $f(x) = \dfrac{2x - 1}{x - 5}$, $6 < f(x) < 8$ **(2)**

....

....

(b) $g(x) = x^2 - 10x + 13$, $0 < x < 11$ **(2)**

> Complete the square and draw a sketch graph of $y = g(x)$.

....

....

....

....

3. The function f is defined by

$$f : x \mapsto \frac{5x + 14}{x^2 + 4x - 12} - \frac{3}{x - 2}, \qquad x \in \mathbb{R}, \ -5 < x < 2$$

(a) Show that $f(x) = \dfrac{2}{x + 6}$ **(3)**

....

....

....

....

(b) Find the range of f. **(2)**

....

....

Inverse functions

> **Guided**

1. The function f is defined by

   $$f : x \mapsto \frac{2x}{3} - 4, \qquad x \in \mathbb{R}, \ -3 < x < 9$$

   (a) Find $f^{-1}(x)$. **(2)**

   $y = \dfrac{2x}{3} - 4$

   $3y = 2x - 12$

   $x = \ldots\ldots\ldots\ldots\ldots\ldots\ldots\ldots\ldots\ldots\ldots\ldots\ldots$

   so the inverse function f^{-1} is $x \mapsto \ldots\ldots\ldots\ldots\ldots\ldots\ldots\ldots\ldots\ldots\ldots$

   (b) Find the domain of f^{-1}. **(2)**

   ....

   ....

2. The function g is defined by

   $$g : x \mapsto \frac{2x + 5}{x - 6}, \qquad x \in \mathbb{R}, \ x > 8$$

   > Write the function in the form $y = \ldots$, then rearrange to make x the subject.

   (a) Find $g^{-1}(x)$. **(3)**

   ....

   ....

   ....

   ....

   (b) Find the domain of g^{-1}. **(2)** > Consider the value of g when $x = 8$ and when $x \to \infty$.

   ....

   ....

3. The function h is defined by

   $$h : x \mapsto \frac{x + 4}{3x + 1}, \qquad x \in \mathbb{R}, \ x > 0$$

   (a) Find $h^{-1}(x)$. **(3)**

   ....

   ....

   ....

   (b) Find the domain of h^{-1}. **(2)** > Consider the value of h when $x = 0$ and when $x \to \infty$.

   ....

   ....

Inverse graphs

Guided 1. The diagram shows part of the curve with equation $y = f(x)$.
The curve intersects the coordinate axes at $(-2, 0)$ and $(0, 4)$.

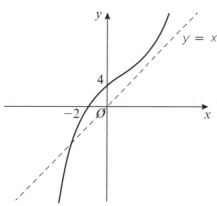

> Draw the line $y = x$, then reflect the graph of $y = f(x)$ in the line $y = x$ to find the graph of the inverse function.

> When reflecting points in the line $y = x$, the x- and y-coordinates swap over.

On the same axes, sketch the curve with equation $y = f^{-1}(x)$. **(2)**

2. The function f has domain $-6 \le x \le 3$ and is linear from $(-6, -4)$ to $(0, -3)$ and from $(0, -3)$ to $(3, 5)$.
A sketch of the graph $y = f(x)$ is shown.
On the same axes, sketch the graph of $y = f^{-1}(x)$.
Show the coordinates of the points corresponding to A and B. **(3)**

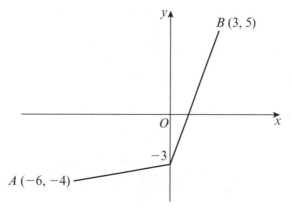

3. The diagrams show the graphs of four functions.
Which of these graphs have inverse functions?
For those that do not have inverse functions, explain why. **(4)**

A 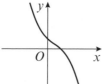   B 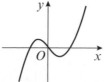   C 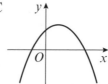   D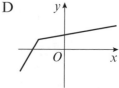

> The inverse of a function exists only if the function is a one-to-one function, which means that the function maps each point in the domain to a single point in the range.

....

....

Modulus

Guided 1. The diagram shows a sketch
of the graph $y = f(x)$.

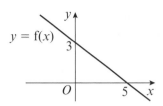

On separate diagrams, sketch the graphs of

(a) $y = |f(x)|$ **(3)**

(b) $y = f(|x|)$ **(3)**

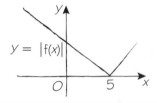

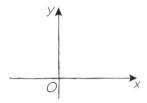

For $y = |f(x)|$, any points on $f(x)$ below
the x-axis are reflected in the x-axis.
All the y-values must be positive.

For $y = f(|x|)$, replace the graph for
values of $x < 0$ with a reflection of the
graph for values of $x > 0$.

Guided 2. For each of the following graphs of $y = f(x)$, sketch the graphs of

(i) $y = |f(x)|$

(ii) $y = f(|x|)$

Label any points of intersection with the axes or any turning points.

(a)

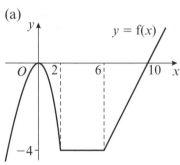

(i)

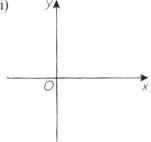

(2)

(ii)

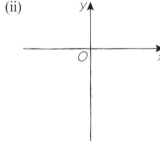

(b)

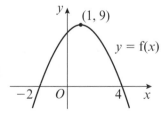

(2)

(c)

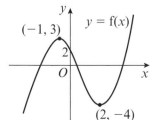

(2)

7

Transformations of graphs

1. The diagram shows the graph of $y = f(x)$, which consists of two line segments meeting at $(-4, 2)$.
Sketch the graph of $y = 3|f(-x)|$ **(3)**

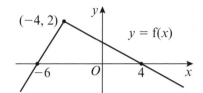

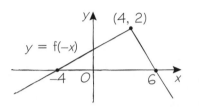 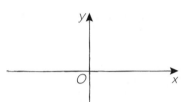

> You can do one step at a time; there are three steps which should be carried out in order.

2. The diagram shows the graph of $y = f(x)$.

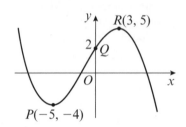

On separate diagrams, sketch the graphs of

(a) $y = f(-x) + 4$ **(3)**

(b) $y = |f(x + 3)|$ **(3)**

On each diagram, mark the positions of the points corresponding to P, Q and R.

> The order in which you do transformations is important.

Modulus equations

Guided   **1.** The function f is defined by

$$f : x \mapsto |7 - 2x|, \qquad x \in \mathbb{R}$$

(a) Sketch the graph with equation $y = f(x)$, showing the points where the graph crosses the axes. **(2)**

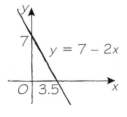 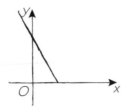

(b) Solve $f(x) = x + 3$ **(2)**

$7 - 2x = x + 3$ $-(7 - 2x) = x + 3$

$x = $ .... $x = $ ....

> You need to solve two equations: a positive argument and a negative argument.

2. Solve $5 - |2x - 4| = 2 - \frac{1}{4}x$ **(5)**

....

....

....

....

3. The function f is defined by $f : x \mapsto |6 - 3x|, \; x \in \mathbb{R}$

(a) Sketch the graph of $y = f(x)$, showing the points where the graph crosses the axes. **(2)**

(b) Explain why $f(x) = -x$ has no solutions. **(1)**

....

(c) Solve $f(x) = x$ **(3)**   > Use your graph to check that the solutions definitely exist.

....

....

....

Sec, cosec and cot

Guided   **1.** Sketch the graph of $y = 2\operatorname{cosec}\left(\frac{1}{2}\theta\right)$

for $-360° < \theta < 360°$ **(2)**

Start with the graph of $y = \operatorname{cosec}\theta$

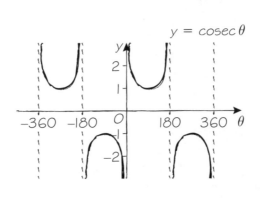

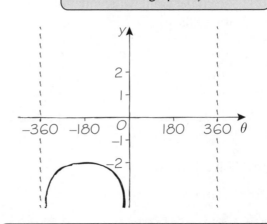

You need to apply two transformations: horizontal stretch scale factor 2, and vertical stretch scale factor 2.

2. Sketch the graph of $y = 2\sec\left(x - \frac{\pi}{2}\right)$ for $0 < x < 2\pi$ **(2)**

You need to be able to work with radians.

Start with $y = \sec x$

3. Sketch the graph of $y = \cot\left(\theta + \frac{\pi}{2}\right) - 1$ for $-\pi < \theta < \pi$ **(2)**

Trig equations 1

Guided

1. Solve for $-180° \leqslant x \leqslant 180°$

$$\sqrt{3}\operatorname{cosec} 2x = 2$$

Give your answers in degrees. **(5)**

$$\operatorname{cosec} 2x = \frac{2}{\sqrt{3}}, \qquad \frac{1}{\sin 2x} = \frac{2}{\sqrt{3}}, \qquad \sin 2x = \frac{\sqrt{3}}{2}$$

$-180° \leqslant x \leqslant 180°$ so $-360° \leqslant 2x \leqslant 360°$

> Remember to double the range for $2x$.

$2x = -300°, \ldots\ldots\ldots, \ldots\ldots\ldots, \ldots\ldots\ldots$

$x = \ldots\ldots\ldots, \ldots\ldots\ldots, \ldots\ldots\ldots, \ldots\ldots\ldots$

2. Solve $\cot 2\theta = \sqrt{3}$ for $0 \leqslant \theta \leqslant \pi$
   Give your answers in terms of π. **(5)**

....

....

....

....

3. Solve $2\sec^2 x = 9\sec x + 5$ for $0° \leqslant x \leqslant 360°$
   Give your answer in degrees to 1 decimal place. **(6)**

....

> Substitute A for $\sec x$ to give an equation you can more easily rearrange as a quadratic and factorise.

....

....

....

....

....

4. Solve $2\operatorname{cosec}\theta - 3 = 5\sin\theta$ for $0° \leqslant \theta \leqslant 360°$
   Give your answers in degrees to 1 decimal place. **(6)**

....

....

....

....

....

....

....

Using trig identities

Guided

1. Prove that $\csc\theta - \sin\theta \equiv \cot\theta\cos\theta$ **(3)**

$$\csc\theta - \sin\theta \equiv \frac{1}{\sin\theta} - \sin\theta \equiv \frac{1 - \sin^2\theta}{\sin\theta} \equiv \dots\dots\dots \equiv \dots\dots\dots\dots$$

> Use $1 - \sin^2\theta = \cos^2\theta$, then rewrite $\cos^2\theta$ as $\cos\theta\cos\theta$ to obtain the result.

> Always start with one side of the identity and use trig manipulation to achieve the expression on the other side.

2. Solve $\sec^2\theta + \tan\theta = 1$ for $0 \leqslant \theta \leqslant 2\pi$
Give your answers in terms of π.   **(5)**

> Use $\sec^2\theta \equiv 1 + \tan^2\theta$

3. Prove that $\sec^2\theta + \csc^2\theta \equiv \sec^2\theta\csc^2\theta$ **(4)**

4. Solve $4\sec^2 2\theta = 5\tan 2\theta + 3\tan^2 2\theta$
for $0° \leqslant \theta \leqslant 180°$
Give your answers in degrees to 1 decimal place.   **(7)**

> Use $\sec^2 2\theta \equiv 1 + \tan^2 2\theta$ and form a quadratic in $\tan 2\theta$. Remember to double the range.

5. Solve $4\cot^2\theta - 3\csc^2\theta = 2\cot\theta$ for $0° \leqslant \theta \leqslant 360°$
Give your answers in degrees to 1 decimal place.   **(7)**

Arcsin, arccos and arctan

1. Write down the value of each of the following in radians.

   (a) $\arcsin\left(\dfrac{1}{2}\right)$ **(1)**   (b) $\arccos\left(\dfrac{-\sqrt{3}}{2}\right)$ **(1)**   (c) $\arctan(-1)$ **(1)**

   ....

> **Guided**

2. The function f is defined by

   $$f : x \mapsto \arccos x + \frac{\pi}{4}, \qquad x \in \mathbb{R}, \ -1 \leqslant x \leqslant 1$$

   (a) Find $f\left(\dfrac{\sqrt{3}}{2}\right)$, giving your answer in terms of π.   **(2)**

   $f\left(\dfrac{\sqrt{3}}{2}\right) = \arccos\left(\dfrac{\sqrt{3}}{2}\right) + \dfrac{\pi}{4} = \dfrac{\pi}{6} + \dfrac{\pi}{4} = \dfrac{\ldots\ldots\ldots}{\ldots\ldots\ldots}$

   (b) Solve the equation $f(x) = \dfrac{7\pi}{12}$, giving your answer as an exact value.   **(3)**

   ....

   ....

   (c) Find $f^{-1}(x)$ and state its domain, and sketch the graph of $y = f^{-1}(x)$, showing the coordinates of the point where the graph crosses the x-axis.   **(6)**

   Let $y = \arccos x + \dfrac{\pi}{4}$

   $y - \dfrac{\pi}{4} = \arccos x$

   $\cos(\ldots\ldots\ldots) = x$

   So $f^{-1}(x) = \ldots\ldots\ldots\ldots$

   Domain of f^{-1} = range of f, so ....

3. The function g is defined by

   $$g : x \mapsto 2\arcsin x + \frac{\pi}{2}, \qquad x \in \mathbb{R}, \ -1 \leqslant x \leqslant 1$$

   The domain of g^{-1} is the range of g.

   Find $g^{-1}(x)$ and state its domain, and sketch the graph of $y = g^{-1}(x)$, showing the coordinates of the point where the graph crosses the x-axis.   **(5)**

   ....

   ....

   ....

   ....

Addition formulae

> **Guided**

1. Find, without using a calculator, an expression in surd form for $\tan 105°$. **(4)**

$$\tan 105° = \tan(60° + 45°) = \frac{\tan 60° + \tan 45°}{1 - \tan 60° \tan 45°} = \frac{\sqrt{3} + 1}{1 - (\sqrt{3})(1)} = \frac{(\sqrt{3} + 1)(1 + \sqrt{3})}{(1 - \sqrt{3})(1 + \sqrt{3})}$$

$$= \frac{\dots\dots\dots\dots}{\dots\dots\dots} = \dots\dots\dots\dots$$

2. Solve, for $0° \leqslant \theta \leqslant 360°$

(a) $\cos(\theta - 30°) = 2\sin\theta$ **(4)**

....

....

....

....

(b) $2\tan\theta = 3\tan(45° - \theta)$ **(5)**

Give your answer in degrees to 1 decimal place.

> Use the addition formula for tan to write this as a quadratic in $\tan\theta$.

....

....

....

....

3. Angle A is acute and angle B is obtuse.

$$\cos A = \frac{5}{7}, \qquad \sin B = \frac{1}{5}$$

(a) Work out the value of $\sin(A - B)$. **(4)**

> First work out the values of $\sin A$ and $\cos B$. Remember that $\cos B$ will be negative.

....

....

....

....

....

(b) Work out the value of $\cos(A - B)$. **(4)**

....

....

....

....

Double angle formulae

> **Guided**

1. Solve, for $-180° \leq \theta \leq 180°$, $1 = 2\sin\theta + 4\cos 2\theta$ **(6)**
Give your answers correct to 1 decimal place.

$1 = 2\sin\theta + 4(1 - 2\sin^2\theta)$

$8\sin^2\theta - 2\sin\theta - 3 = 0$

$(\ldots\ldots\ldots\ldots)(\ldots\ldots\ldots\ldots) = 0$

$\sin\theta = \ldots\ldots\ldots\ldots$ or $\sin\theta = \ldots\ldots\ldots\ldots$

$\theta = \ldots\ldots\ldots\ldots, \ldots\ldots\ldots\ldots$ $\theta = \ldots\ldots\ldots\ldots, \ldots\ldots\ldots\ldots$

> The single term in the equation is $\sin\theta$, so choose the appropriate version of the $\cos 2\theta$ formula to produce a quadratic in $\sin\theta$. Factorise and solve.

2. θ is an obtuse angle and $\sin\theta = \frac{2}{3}$

Use the double angle formulae to work out the values of

> $\cos\theta$ and $\tan\theta$ will be negative.

(a) $\sin 2\theta$ **(2)**   (b) $\cos 2\theta$ **(2)**   (c) $\tan 2\theta$ **(2)**

....

....

....

....

....

3. Prove that $\dfrac{\sin 3A}{\sin A} + \dfrac{\cos 3A}{\cos A} = 4\cos 2A$ **(5)**

> Start with the LHS and write it as a fraction with common denominator $\sin A \cos A$. Use the addition formulae and double angle formulae. Remember that $\sin 2\theta = 2\sin\theta\cos\theta$ also implies that $\sin 4\theta = 2\sin 2\theta\cos 2\theta$.

....

....

....

....

....

....

4. Solve, for $0° \leq x \leq 360°$,

$\cos 2x - 3 = 5\cos x$

Give your answers correct to 1 decimal place.   **(6)**

> First use a double angle formula for $\cos 2x$, then solve a quadratic in $\cos x$. You will need to use the quadratic equation formula.

....

....

....

....

....

....

$a \cos \theta \pm b \sin \theta$

Guided

1. (a) Express $4 \cos \theta + 2 \sin \theta$ in the form $R \cos (\theta - \alpha)$ where $R > 0$ and $0° < \alpha < 90°$ **(4)**

$4 \cos \theta + 2 \sin \theta \equiv R(\cos \theta \cos \alpha + \sin \theta \sin \alpha)$

$R \cos \alpha = 4, \qquad R \sin \alpha = 2, \qquad \tan \alpha = \frac{1}{2}, \qquad \alpha = \ldots\ldots\ldots$

$R = \sqrt{4^2 + \ldots\ldots\ldots} = \sqrt{\ldots\ldots\ldots}$

So $4 \cos \theta + 2 \sin \theta \equiv \sqrt{\ldots\ldots\ldots} \cos(\theta - \ldots\ldots\ldots)$

(b) Hence, or otherwise, solve

$$4 \cos \theta + 2 \sin \theta = 1,$$

for $0° \leqslant \theta \leqslant 360°$

Give your answers correct to 1 decimal place.   **(5)**

> Use the result from part (a).
> Look at the range of values of θ
> and check that you have found
> all the solutions in the range.

(c) State the maximum and minimum values of $4 \cos \theta + 2 \sin \theta$ **(2)**

2. (a) Express $5 \sin x - 8 \cos x$ in the form

$$R \sin (x - \alpha)$$

where $R > 0$ and $0 < \alpha < \dfrac{\pi}{2}$ **(4)**

(b) Hence, or otherwise, solve

$$5 \sin x - 8 \cos x = 6, \qquad 0 \leqslant x \leqslant 2\pi$$

giving your answers to 2 decimal places.   **(5)**

> Remember to put your calculator
> into radians mode. Don't round
> prematurely and check that you
> have found all the solutions in
> the specified range.

Trig modelling

1. (a) Express $6\cos\theta + 8\sin\theta$ in the form $R\cos(\theta - \alpha)$ where $R > 0$ and $0 < \alpha < \frac{\pi}{2}$ **(4)**

$6\cos\theta + 8\sin\theta \equiv R(\cos\theta\cos\alpha + \sin\theta\sin\alpha)$

$R\cos\alpha = 6, \qquad R\sin\alpha = 8, \qquad \tan\alpha = \frac{8}{6}, \qquad \alpha = \ldots\ldots\ldots$

$R = \sqrt{\ldots\ldots\ldots + \ldots\ldots\ldots} = \sqrt{\ldots\ldots\ldots} = \ldots\ldots\ldots$

So $6\cos\theta + 8\sin\theta \equiv \ldots\ldots\ldots \cos(\theta - \ldots\ldots\ldots)$

> Don't round any values until the end of the calculation. Work in radians and show at least 4 decimal places in your answers.

(b) (i) Find the maximum value of $6\cos\theta + 8\sin\theta$

....

  (ii) Find the smallest value of θ for which this maximum occurs. **(2)**

....

The temperature, in °C, of a storage depot, is modelled by the equation

$$f(t) = 14 + 6\cos(0.25t) + 8\sin(0.25t)$$

where t is the time in hours after midday, $0 \leqslant t \leqslant 24$.

(c) Calculate the maximum and minimum temperatures predicted by this model and find the values of t for which the maximum and minimum temperatures occur. **(4)**

....

....

....

....

(d) Find the times, to the nearest minute, when the temperature of the depot is 12 °C. **(6)**

> t is the time in hours. Remember to convert to hours and minutes, then round to the nearest minute.

....

....

....

....

....

....

(e) Sketch a graph of $f(t) = 14 + 6\cos(0.25t) + 8\sin(0.25t)$, showing clearly the coordinates of the maximum and minimum points. **(3)**

Exponential functions

> **Guided**

1. (a) The functions f and g are defined by

$$f : x \mapsto 2x + \ln 3, \qquad x \in \mathbb{R}$$

$$g : x \mapsto e^x, \qquad x \in \mathbb{R}$$

Find gf and state its range. **(3)**

$gf(x) = e^{(2x + \ln 3)} = e^{2x} \times e^{\ln 3} = \dots$

Range of $gf(x)$ is ....

> The range of gf is the same as the range of g.

(b) Find $(gf)^{-1}$ and sketch a graph showing $gf(x)$ and $(gf)^{-1}(x)$. **(5)**

> $(gf)^{-1}$ is the inverse function of gf. Start with $y = gf(x)$ and rearrange to make x the subject. You will need to take logs of both sides to get x on its own.

....

....

....

....

....

....

(c) Find f^{-1}, stating its domain. **(3)**

....

....

....

2. The function f is defined by

$$f : x \mapsto \ln(6 - 4x), \qquad x \in \mathbb{R}, \ x < 1.5$$

(a) Find f^{-1}, stating its domain and range. **(5)**

....

....

....

....

(b) Sketch the graph of $y = f^{-1}(x)$, stating the coordinates of the points of intersection with the x- and y-axes. **(4)**

Exponential equations

> **Guided**

1. Solve $\ln(5x + 24) = 2\ln(x + 2)$ **(4)**

   $\ln(5x + 24) = \ln(x + 2)^2$

   $5x + 24 = (x + 2)^2$

   > Rearrange as a quadratic in x, factorise and solve. Check the validity of your answers.

   ....

   ....

2. Solve $\ln(x - 3) + \ln(x - 2) = \ln(2x + 24)$ **(5)**

   ....

   ....

   ....

3. Find the exact solutions of

   $$e^{3x} + 2e^x = 3e^{2x}$$ **(5)**

   > Write e^{3x} as $(e^x)^3$ and e^{2x} as $(e^x)^2$. Take all terms to the LHS then take out e^x as a common factor. You will have a quadratic in e^x as the other factor.

   ....

   ....

   ....

   ....

4. Solve $3^x e^{2x-1} = 5$

   Give your answers in the form $\dfrac{\ln a + b}{\ln c + d}$

   where a, b, c and d are integers. **(5)**

   > Take logs of both sides, then use the laws of logs to simplify the LHS. Group the x-terms together.

   ....

   ....

   ....

5. The function f is defined by

   $$f : x \mapsto \frac{5x^2 - 13x - 6}{x^2 - 9}, \qquad x > 3$$

   (a) Show that $f(x) = \dfrac{5x + 2}{x + 3}$ **(3)**

   ....

   ....

   (b) Hence, or otherwise, solve the equation $\ln(5x^2 - 13x - 6) = 2 + \ln(x^2 - 9)$, $x > 3$, giving your answer in terms of e. **(4)**

   ....

   ....

   ....

Exponential modelling

> **Guided**

1. A heated metal bar is put in a liquid. The temperature of the bar, $T°C$, at time t minutes is modelled by the equation

   $$T = 350\,e^{-0.08t} + 20, \qquad t \geq 0$$

   (a) Write down the temperature of the bar as it enters the liquid.   **(1)**

   ....

   (b) Find t when $T = 280$, giving your answers to 3 s.f.   **(4)**

   $280 = 350\,e^{-0.08t} + 20$

   $\dfrac{260}{350} = e^{-0.08t} \qquad$ so $\qquad \ln\left(\dfrac{260}{350}\right) = -0.08t$

   ....

   (c) Find the rate at which the temperature is decreasing at time $t = 40$.
   Give your answer in °C/minute to 3 s.f.   **(3)**

   > $\dfrac{dT}{dt} = -28e^{-0.08t}$ °C/min
   > gives the rate of change.

   ....

   ....

   (d) Explain why the temperature can never fall to 18 °C.   **(1)**

   ....

2. A sample of radioactive material decays according to the formula $N = 60\,e^{-kt}$ where N is the number of grams of the material, t is the time, in years, and k is a positive constant.

   (a) What was the initial mass of the sample?   **(1)**

   ....

   After 88 years, the sample has lost half its mass.

   (b) Find the value of k to 3 s.f.   **(4)**

   ....

   ....

   (c) How many grams of material are there after 120 years?   **(2)**

   ....

   (d) Sketch a graph of N against t.   **(2)**

The chain rule

> **Guided**

1. Find $\dfrac{dy}{dx}$ for each of the following.

(a) $y = (5 - 3x)^7$ **(2)**

$u = 5 - 3x$ $y = u^7$

$\dfrac{du}{dx} = -3$ $\dfrac{dy}{du} = 7u^6$

$\dfrac{dy}{dx} = \dfrac{dy}{du} \times \dfrac{du}{dx} = \dots\dots \times \dots\dots = \dots\dots\dots\dots$

> Remember to write your final answer in terms of x, not u.

(b) $y = (4 - x^2)^{-4}$ **(2)**

$\dfrac{dy}{dx} = -4 \times (4 - x^2)^{-5} \times (\dots\dots) = \dots\dots$

> The missing term is the derivative of $(4 - x^2)$. Always simplify your answer.

(c) $y = (1 + 6x)^{\frac{3}{2}}$ **(3)**

....

....

(d) $y = (4x + 5)^{\frac{1}{2}}$ **(3)**

....

....

(e) $y = \dfrac{1}{\sqrt{3 - 2x^2}}$ **(4)**

....

....

....

(f) $y = \dfrac{2}{\sqrt[3]{3x^2 + 4}}$ **(4)**

> $y = 2(3x^2 + 4)^{-\frac{1}{3}}$

....

....

2. A curve has equation $y^3 + 3y^2 - 4y - 5 = x$

(a) Find $\dfrac{dy}{dx}$ in terms of y.   **(3)**

> Use $\dfrac{dy}{dx} = \dfrac{1}{\frac{dx}{dy}}$

....

....

(b) Find the gradient of the curve at the point $(7, -2)$.   **(2)**

....

> $\dfrac{dy}{dx}$ is given in terms of y so substitute the y-coordinate to find the gradient.

3. $f(x) = (4\sqrt{x} + 3)^3$

Find $f'(x)$.   **(2)**   Simplify your answer as much as possible.

....

....

Derivatives to learn

> Guided

1. Differentiate with respect to x

   (a) $y = \sin^3 x$ **(2)**

   $u = \sin x \qquad y = u^3$

   $\dfrac{du}{dx} = \cos x \qquad \dfrac{dy}{du} = 3u^2$

   $\dfrac{dy}{dx} = \dfrac{dy}{du} \times \dfrac{du}{dx} = \text{............} \times \text{............} = \text{......................}$

   (b) $y = \cos(5 - 4x)$ **(2)**

   ....

   ....

   (c) $y = e^{x^2+1}$ **(2)**

   (d) $y = \ln(x^3 + 2)$ **(2)**

   > When using the chain rule, don't forget to differentiate the expression inside the brackets.

   ....

   ....

2. Work out $\dfrac{dy}{dx}$ for each of the following.

   (a) $y = \cos^4 x$ **(2)**

   ....

   ....

   (b) $y = \sin 2x + \cos 3x$ **(2)**

   ....

   ....

   (c) $y = 2x^3 - e^{5-3x}$ **(3)**

   (d) $y = \ln(\cos 2x)$ **(2)**   > Simplify your answer.

   ....

   ....

3 (a) Given that $y = \operatorname{cosec} x$, show that $\dfrac{dy}{dx} = -\operatorname{cosec} x \times \cot x$ **(3)**

   ....

   ....

   ....

   (b) Hence, or otherwise, differentiate $\ln(\operatorname{cosec} x)$ with respect to x. **(2)**

   ....

   ....

   ....

   ....

   ....

The product rule

Guided

1. Differentiate with respect to x

(a) $y = (x^2 + 3)(2x^3 - 1)$ **(3)**

$$u = x^2 + 3 \qquad v = 2x^3 - 1$$

$$\frac{du}{dx} = 2x \qquad \frac{dv}{dx} = 6x^2$$

> Expand the brackets and simplify.

$$\frac{dy}{dx} = u\frac{dv}{dx} + v\frac{du}{dx} = (x^2 + 3)(6x^2) + (2x^3 - 1)(2x) = \dots\dots\dots\dots\dots\dots\dots$$

....

(b) $y = x^4 \sin x$ **(3)**

$$u = x^4 \qquad v = \sin x$$

$$\frac{du}{dx} = 4x^3 \qquad \frac{dv}{dx} = \dots\dots\dots$$

$$\frac{dy}{dx} = x^4 \dots\dots\dots + \sin x(4x^3)$$

(c) $y = 2x^2 e^{2x}$ **(3)**

....

....

(d) $y = (x^3 - 4x)\ln 2x$ **(3)**

....

....

2. (a) Given that $h(x) = e^{4x}\sec x$, find $h'(x)$. **(4)**

....

....

(b) Solve, for $-\dfrac{\pi}{2} < x < \dfrac{\pi}{2}$, the equation $h'(x) = 0$.
Give your answer correct to 3 s.f. **(4)**

> Make sure your calculator is in radians mode.

....

....

3. If $y = x^2\sqrt{1 + x^2}$, show that $\dfrac{dy}{dx} = \dfrac{3x^3 + 2x}{\sqrt{1 + x^2}}$ **(5)**

> Use the product rule, then use $\sqrt{1 + x^2}$ as a common denominator.

....

....

....

4. A curve has equation $y = (x^2 - 2x - 1)e^{2x}$
Find the coordinates of the turning points on the curve. **(6)**

....

....

....

The quotient rule

> **Guided**

1. Differentiate with respect to x, simplifying your answers.

(a) $y = \dfrac{x^2}{2x + 1}$ **(4)**

$u = x^2 \qquad\qquad v = 2x + 1$

$\dfrac{du}{dx} = 2x \qquad\qquad \dfrac{dv}{dx} = 2$

$\dfrac{dy}{dx} = \dfrac{v\dfrac{du}{dx} - u\dfrac{dv}{dx}}{v^2} = \dfrac{(2x + 1)(2x) - (x^2)(2)}{(2x + 1)^2} = \dfrac{\text{.................}}{(2x + 1)^2}$

(b) $y = \dfrac{x^3}{\sqrt{1 - 2x^2}}, \qquad x^2 \neq \tfrac{1}{2}$ **(5)**

$u = x^3 \qquad\qquad\qquad v = (1 - 2x^2)^{\frac{1}{2}}$

$\dfrac{du}{dx} = 3x^2 \qquad\qquad \dfrac{dv}{dx} = \tfrac{1}{2}(1 - 2x^2)^{\frac{-1}{2}}(-4x) = (-2x)(1 - 2x^2)^{\frac{-1}{2}}$

....

....

2. Differentiate with respect to x, simplifying your answers.

(a) $f(x) = \dfrac{x^2 + x - 1}{1 - x^2}, \qquad x \neq \pm 1$ **(4)** (b) $f(x) = \dfrac{4x}{\sqrt{3x^2 + 1}}$ **(5)**

....

....

....

....

....

3. Differentiate with respect to x, simplifying your answers.

(a) $\dfrac{\ln 2x}{3x}$ **(4)** (b) $\dfrac{\cos 3x}{1 + \sin 3x}$ **(5)** (c) $\dfrac{2\sin x - 3\cos x}{e^{2x}}$ **(5)**

....

....

....

....

....

....

....

Differentiation and graphs

1. A curve C has equation $y = 3x^2 \ln x$

   (a) Find $\dfrac{dy}{dx}$ $\dfrac{dy}{dx} = (3x^2)\left(\dfrac{1}{x}\right) + (\ln x)(6x) = 3x + 6x\ln x$ **(3)**

   (b) Find an equation of the tangent to C at the point where $x = e$ **(3)**

When $x = e$, $y = 3e^2 \ln e = $ ....

When $x = e$, $\dfrac{dy}{dx} = 3e + 6e\ln e = $ .... $ = $ ....

Equation of tangent is ....

> Use $y - y_1 = m(x - x_1)$

   (c) The tangent to C when $x = e$ intersects the x-axis at point A.
      Find the exact values of the coordinates of A.   **(2)**

....

....

2. A curve C has equation $y = e^{2x} \cos 3x$, $-\dfrac{\pi}{2} < x < \dfrac{\pi}{2}$

> A curve has a turning point at $\dfrac{dy}{dx} = 0$.

   (a) Show that turning points on C occur when $\tan 3x = \dfrac{2}{3}$ **(4)**

....

....

....

   (b) Find an equation of the normal to C at the point where $x = 0$.
      Give your answer in the form $ax + by = c$, where a, b and c are integers.   **(4)**

....

....

3. A curve C has equation $y = \dfrac{4x}{x^2 + 4}$

   (a) Find the maximum and minimum points on the curve.   **(7)**

....

....

....

....

   (b) Find an equation of the tangent to C at the point where $x = 1$.
      Give your answer in the form $ax + by + c = 0$, where a, b and c are integers.   **(3)**

....

....

Iteration

Guided

1. $f(x) = e^{-x} - 3 + 2\sqrt{x}$

(a) Show that the equation $f(x) = 0$ has a root, α, between 2 and 3. **(2)**

> You need to show that f(2) is negative and f(3) is positive, and write a conclusion.

$f(2) = e^{-2} - 3 + 2\sqrt{2} = \ldots\ldots\ldots$

$f(3) = e^{-3} - 3 + 2\sqrt{3} = \ldots\ldots\ldots$

Hence there is a root between 2 and 3 because there is a change of sign.

(b) Show that the equation $f(x) = 0$ can be written as $x = \frac{1}{4}(3 - e^{-x})^2$ **(3)**

$e^{-x} - 3 + 2\sqrt{x} = 0, \qquad 2\sqrt{x} = 3 - e^{-x}, \qquad \sqrt{x} = \ldots\ldots\ldots\ldots, \qquad x = \ldots\ldots\ldots\ldots$

(c) Starting with $x_0 = 2.1$, use the iterative formula $x_{n+1} = \frac{1}{4}(3 - e^{-x_n})^2$ to calculate x_1, x_2, and x_3, giving your answers correct to 4 decimal places. **(3)**

$x_1 = \frac{1}{4}(3 - e^{-2.1})^2 = \ldots\ldots\ldots\ldots\ldots\ldots\ldots\ldots\ldots$

> Take great care using your calculator and show at least 6 d.p. before you round to 4 d.p.

$x_2 = \ldots\ldots\ldots\ldots\ldots\ldots\ldots\ldots\ldots$

$x_3 = \ldots\ldots\ldots\ldots\ldots\ldots\ldots\ldots\ldots$

(d) By choosing a suitable interval, show that $\alpha = 2.064$, correct to 3 decimal places. **(3)**

....

> Substitute 2.0635 and 2.0645 into the expression for f(x) and look for a change of sign. Write 'change of sign' and write down the interval that contains the root.

....

....

....

....

....

2. $f(x) = x^3 + 3x^2 - 4x - 2$

(a) Show that the equation $f(x) = 0$ has a root, α, between 1 and 2. **(2)**

....

(b) Show that the equation $f(x) = 0$ can be written as $x = \sqrt{\dfrac{4x + 2}{x + 3}}$, $x \neq -3$ **(2)**

....

(c) Starting with $x_0 = 1.5$, use the iterative formula $x_{n+1} = \sqrt{\dfrac{4x_n + 2}{x_n + 3}}$ to calculate x_1, x_2, and x_3, giving your answers correct to 4 decimal places. **(3)**

....

(d) By choosing a suitable interval, show that $\alpha = 1.2924$, to 4 decimal places. **(3)**

....

....

You are the examiner!

Checking your work is a key skill for A2 Maths. Have a look at pages 27 and 28 of the *Revision Guide*, then practise with these questions. There are full worked solutions on page 63.

1. Express $\dfrac{5x}{x^2 - 2x - 24} - \dfrac{3}{x - 6} + \dfrac{2}{x + 4}$ as a single fraction in its simplest form. **(4)**

> Always use brackets – it helps you to avoid sign errors.

2. (a) Express $3\sin\theta + 7\cos\theta$ in the form $R\sin(\theta + \alpha)$ where $R > 0$ and $0 < \alpha < 90°$.
Find the exact value of R and the value of θ correct to 1 decimal place. **(4)**

> 'Find the exact value' means write R as a surd.

(b) Hence, or otherwise, solve the equation $3\sin\theta + 7\cos\theta = 4$ for $0° \leqslant \theta \leqslant 360°$.
Give your answers correct to 1 decimal place. **(5)**

> Make sure you adjust the range of values for $(\theta + \alpha)$ when you look for solutions.

You are the examiner!

Checking your work is a key skill for A2 Maths. Have a look at pages 27 and 28 of the *Revision Guide*, then practise with these questions. There are full worked solutions on pages 63 and 64.

3. Given that $f(x) = \ln(2x + 3)$

   (a) find $f^{-1}(x)$ **(3)**

....

....

....

   (b) sketch the graph of $y = f^{-1}(x)$, stating the coordinates of the points of
intersection with the x- and y-axes and the equations of any asymptotes.   **(4)**

> $f^{-1}(x)$ is an exponential function, so you will need to draw and label the asymptote, as well as labelling points of intersection with the axes.

4. Given that $y = \dfrac{4 - 3x}{e^{2x}},\ x > \tfrac{4}{3},$ find $\dfrac{dy}{dx}.$ **(3)**

> Use the quotient rule and make sure you define, and use, u and v correctly.

....

....

....

....

5. (a) Solve the equation $\ln(3x - 2) + \ln(x - 2) = 2\ln(x + 2),\ x > 2$ **(5)**

> Use the rules of logarithms on both sides of the equation.

....

....

....

   (b) Solve the equation $2e^x + 5 = 3e^{-x}$ **(5)**

> Multiply through by e^x to give a quadratic in e^x.

....

....

....

Partial fractions

Guided 1. Express in partial fractions $\dfrac{3x - 10}{(x - 2)(x - 4)}$ **(3)**

> You can use the cover-up rule when there are no repeated factors.

$\dfrac{3x - 10}{(x - 2)(x - 4)} = \dfrac{A}{x - 2} + \dfrac{B}{x - 4}$

When $x = 2$, work out $\dfrac{3x - 10}{x - 4}$: $A = \dfrac{3(2) - 10}{2 - 4} = \dfrac{-4}{-2} = 2$

When $x = 4$, work out $\dfrac{3x - 10}{x - 2}$: $B = \dfrac{3(\ldots\ldots) - 10}{(\ldots\ldots - 2)} = \dfrac{\ldots\ldots}{\ldots\ldots} = \ldots\ldots$

So $\dfrac{3x - 10}{(x - 2)(x - 4)} = \dfrac{2}{x - 2} + \dfrac{\ldots\ldots}{x - 4}$

2. Express $\dfrac{10x - 1}{(2x + 1)(4x - 1)}$ in partial fractions.   **(3)**

3. Express $\dfrac{6 - x}{x^3 - x^2 - 6x}$ in partial fractions.   **(5)**

> First, fully factorise the denominator.

Guided 4. $\dfrac{x^2 - 13}{(x - 1)^2(x + 2)} = \dfrac{A}{x - 1} + \dfrac{B}{(x - 1)^2} + \dfrac{C}{x + 2}$

Find the values of the constants A, B and C.   **(4)**

> There is a repeated factor so you will need to substitute values of x and/or compare coefficients.

$x^2 - 13 = A(x - 1)(x + 2) + B(x + 2) + C(x - 1)^2$

When $x = -2$, $-9 = 9C$, so $\underline{C = -1}$

5. $\dfrac{x^3 - 3x^2 + 1}{x^2 - x - 2} = Ax + B + \dfrac{C}{x + 1} + \dfrac{D}{x - 2}$

Find the values of the constants A, B, C and D.   **(5)**

> It is easier to manage this if you write
> $x^3 - 3x^2 + 1 = (Ax + B)(x + 1)(x - 2) + C(x - 2) + D(x + 1)$

Parametric equations

Guided   **1.** The curve C has parametric equations

$$x = 1 - t, \qquad y = t^2 - 4$$

Find a cartesian equation of the curve. **(3)**

$t = 1 - x$, so $t^2 = (1 - x)^2 = $ ....

$y = t^2 - 4 = $ .... $- 4$ i.e. $y = $ ....

2. Work out cartesian equations for these curves.

   (a) $x = e^{2t}, \qquad y = 3e^{-t}$ **(3)**   (b) $x = t^2 - 1, \qquad y = t^4 + 1$ **(3)**

   (c) $x = \sec\theta, \qquad y = 5\tan\theta$ **(3)**   (d) $x = 1 + \cos\theta, \qquad y = 1 - 2\sin\theta$ **(3)**

3. A curve C has parametric equations

$$x = t - 1, \qquad y = \frac{1}{t}$$

A line L has equation $3x - 2y = 2$

Find the points of intersection of C and L. **(6)**

> Substitute the expressions for x and y in terms of t into the equation of the line, then solve the resulting quadratic in t. Don't forget to state the (x, y) solutions.

4. The curve C has parametric equations

$$x = 2\cos\theta + \sin\theta, \qquad y = \cos\theta - \sin\theta$$

Find a cartesian equation of the curve. **(3)**

> Solve simultaneous equations for $\sin\theta$ and $\cos\theta$, then use $\sin^2\theta + \cos^2\theta = 1$.

Parametric differentiation

Guided **1.** A curve C has parametric equations

$$x = 4\cos t, \qquad y = 3\sin t, \qquad 0 < t < \frac{\pi}{2}$$

(a) Find $\dfrac{dy}{dx}$ in terms of t. **(4)**

$$\frac{dx}{dt} = -4\sin t, \qquad \frac{dy}{dt} = \dots\dots\dots\dots$$

$$\frac{dy}{dx} = \frac{dy}{dt} \div \frac{dx}{dt} = \frac{\dots\dots\dots\dots\dots}{\dots\dots\dots\dots\dots} = \dots\dots\dots\dots\dots\dots$$

(b) Find an equation of the tangent to C when $t = \dfrac{\pi}{3}$. **(5)**

....

....

....

....

2. A curve C has parametric equations

$$x = 4e^{-2t} - 3, \qquad y = 2e^{2t} - 5$$

(a) Find $\dfrac{dy}{dx}$ in terms of t. **(3)**

....

....

The point P, where $t = \ln 2$, lies on curve C.

(b) Find the gradient of the tangent at P. **(1)**

....

(c) Find the coordinates of P. **(2)**

....

....

The normal at P crosses the x-axis at Q.

(d) Find the coordinates of Q. **(4)**

....

....

(e) Find a cartesian equation of C in the form $xy + ay + bx = k$ where a, b and k are integers. **(5)**

....

....

....

....

The binomial series

> **Guided**

1. $f(x) = (1 + 4x)^{\frac{3}{2}}$, $|x| < \frac{1}{4}$

   Find the binomial expansion of $f(x)$, in ascending powers of x, up to and including the term in x^3. Give each coefficient in its simplest form. **(5)**

   > Take great care working out the coefficients. For x^2:
   > $$\frac{\left(\frac{3}{2}\right)\left(\frac{1}{2}\right)(4^2)}{1 \times 2} = \frac{\frac{3}{2} \times \frac{1}{2} \times 16}{1 \times 2} = \frac{3 \times 4}{1 \times 2} = 6$$

   $$(1 + 4x)^{\frac{3}{2}} = 1 + \left(\tfrac{3}{2}\right)(4x) + \frac{\left(\frac{3}{2}\right)\left(\frac{1}{2}\right)(4x)^2}{1 \times 2} + \frac{\dots\dots\dots(4x)^3}{1 \times 2 \times 3} + \dots\dots$$

   $$= 1 + 6x + 6x^2 \dots\dots\dots\dots\dots$$

> **Guided**

2. Find the series expansion of $\sqrt{4 - x}$, $|x| < 4$, up to and including the term in x^3, simplifying each term. **(7)**

   > Don't forget to multiply all the terms by 2.

   $$\sqrt{4 - x} = (4 - x)^{\frac{1}{2}} = \left[4\left(1 - \frac{x}{4}\right)\right]^{\frac{1}{2}} = 4^{\frac{1}{2}}\left(1 - \frac{x}{4}\right)^{\frac{1}{2}} = 2\left(1 - \frac{x}{4}\right)^{\frac{1}{2}}$$

   $$2\left(1 - \frac{x}{4}\right)^{\frac{1}{2}} = 2\left[1 + \left(\tfrac{1}{2}\right)\left(\frac{-x}{4}\right) + \frac{(\dots\dots)(\dots\dots)\left(\frac{-x}{4}\right)^2}{1 \times 2} + \frac{\dots\dots\dots\dots}{\dots\dots\dots\dots} + \dots\dots\right]$$

   $$\dots\dots\dots\dots\dots\dots\dots\dots\dots\dots\dots\dots\dots\dots\dots\dots\dots\dots\dots\dots\dots\dots$$

3. (a) Expand $\dfrac{1}{\sqrt{4 + 5x}}$, $|x| < 0.8$, in ascending powers of x up to and including the term in x^3, simplifying each term. **(5)**

   (b) Use your expansion, with a suitable value of x, to work out an approximate value of $\dfrac{1}{\sqrt{3.9}}$, giving your answer to 5 decimal places. State the value of x which you use in your expansion, and show all your working. **(3)**

   (c) Find the first three terms in the series expansion of $\dfrac{x + 4}{\sqrt{4 + 5x}}$, simplifying each term. **(3)**

Implicit differentiation

1. A curve C has equation $x^2 + 4y - y^2 = 7$

   Find an equation of the normal to C at the point $(-2, 3)$.

   Give your answers in the form $ax + by + c = 0$, where a, b and c are integers. **(7)**

   Differentiating implicitly, $2x + 4\dfrac{dy}{dx} - 2y\dfrac{dy}{dx} = 0$

   $2x = (2y - 4)\dfrac{dy}{dx}$

   $\dfrac{dy}{dx} = \dfrac{\text{..................}}{\text{..................}} = \dfrac{\text{............}}{\text{............}}$

   When $x = -2$ and $y = 3$, $\dfrac{dy}{dx} = \text{..................}$ So gradient of normal $= \text{..................}$

   Equation of normal is $y - 3 = \text{............} (x - {-2})$

   > Remember to write your final equation in the correct form.

   ....

2. A curve C has equation

   $$3x^2 - y^2 + 3xy + 28 = 0$$

   Find the coordinates of the points

   on C where $\dfrac{dy}{dx} = 0$ **(7)**

   > Differentiate implicitly and equate to zero. This will give a connection between x and y. Use this to substitute into the equation for C and solve an equation in one variable.

   ....

   ....

   ....

   ....

   ....

3. A curve C has equation $\cos 2y + y\,e^{2x} = \pi$

   The point $P(\ln 2, \frac{\pi}{4})$ lies on C.

   (a) Find an expression for $\dfrac{dy}{dx}$ in terms of x and y. **(5)**

   ....

   ....

   ....

   ....

   (b) Find the exact value of the gradient of the tangent at P. **(2)**

   ....

   ....

   (c) The tangent at P crosses the x-axis at Q. Find the exact value of the x-coordinate of Q. **(2)**

   ....

   ....

Differentiating a^x

Guided 1. Given that $y = 4^{2x}$, find $\dfrac{dy}{dx}$ **(2)**

> You can use the chain rule, and the rule $\dfrac{d}{dx}(a^x) = a^x \ln x$.

Let $u = 2x$ $\qquad y = 4^u$

$\dfrac{du}{dx} = 2$ $\qquad \dfrac{dy}{du} = 4^u \ln 4$

$\dfrac{dy}{dx} = \dfrac{dy}{du} \times \dfrac{du}{dx} = \text{..................} = \text{...................}$

2. Differentiate, with respect to x

   (a) $y = 3^{\tan x}$ **(2)**   (b) $y = x^3\, 6^{2x}$ **(3)**   (c) $y = 5^{x^2} \ln 2x$ **(3)**

3. A curve has equation $2xy + 3^y = 9$
Find an equation of the normal to
the curve at the point $(0, 2)$. **(6)**

> The gradient of the normal is $\dfrac{-1}{\frac{dy}{dx}}$

4. A curve C has equation $4^x - 3y^2 = 4xy$

   (a) Find the coordinates of the two points on C where $x = 2$. **(3)**

   (b) Find the exact value of $\dfrac{dy}{dx}$ at each of the points where $x = 2$. **(7)**

Rates of change

Guided 1. Water is poured into a conical vessel at a rate of $8\,cm^3\,s^{-1}$.

   After t seconds, the volume $V\,cm^3$ of water is given by $V = \frac{2}{9}\pi x^3$, where x is the depth of the water.

   Find, in terms of π, the rate at which the water is rising when $x = 6$. **(5)**

   $\dfrac{dV}{dt} = 8, \qquad V = \frac{2}{9}\pi x^3, \qquad so\ \dfrac{dV}{dx} = \dots\dots$

   Rate required $= \dfrac{dx}{dt} = \dfrac{dx}{dV} \times \dfrac{dV}{dt} = \dfrac{dV}{dt} \div \dfrac{dV}{dx} = 8 \div \dots\dots = \dots\dots$

   > Use the chain rule.

   When $x = 6, \dfrac{dx}{dt} = \dots\dots\ cm\,s^{-1}$

2. The volume of a spherical bubble is increasing at the rate of $5\,cm^3\,s^{-1}$.
   Find the rate at which the radius is increasing when the radius is $4\,cm$. Give your answer correct to 3 significant figures. **(4)**

   > Volume of sphere $= \frac{4}{3}\pi r^3$

   ....

   ....

   ....

   ....

3. A horizontal trough is $4\,m$ long and $1\,m$ deep.
   Its cross-section is an isosceles triangle of
   width $1.5\,m$. When the depth of the water is $h\,m$,
   the volume is $V\,m^3$ and the width of the
   cross-section of the water is $x\,m$.

   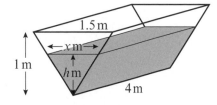

   (a) Show that $V = 3h^2$. **(3)**

   ....

   ....

   ....

   ....

   (b) Water runs into the trough at the rate of $0.03\,m^3\,s^{-1}$.
       Find the rate at which the water level is rising at the instant when the water has
       been running for 25 seconds. **(8)**

   ....

   ....

   ....

   ....

   ....

   ....

Integrals to learn

> **Guided**   1. Find

> Check your answer by differentiating to ensure you have the correct coefficient.

    (a) $\int 6\cos 2x\,dx$ **(2)**

        = ... $\sin 2x + c$

    (b) $\int \sec^2 \frac{x}{2}\,dx$ **(2)**

        = ...

    (c) $\int \frac{1}{2}\sin 3x\,dx$ **(2)**

        = ... $\cos 3x + c$

    (d) $\int \csc 2\theta \cot 2\theta\,d\theta$ **(2)**

        = ...

2. Find

    (a) $\int -12\sin 4x\,dx$ **(2)**

    (b) $\int 4\cos \frac{\theta}{2}\,d\theta$ **(2)**

    (c) $\int 5\sin 10x\,dx$ **(2)**

    (d) $\int \frac{1}{2}\cos (3x + 5)\,dx$ **(2)**

3. Find

    (a) $\int 2e^{3x}\,dx$ **(2)**

    (b) $\int \frac{3}{5}e^{\frac{\theta}{4}}\,d\theta$ **(2)**

    (c) $\int \frac{5}{2x - 1}\,dx$ **(2)**

    (d) $\int \frac{2}{1 + 8x}\,dx$ **(2)**

4. Find

    (a) $\int \sqrt{x + 6}\,dx$ **(2)**

    (b) $\int \frac{1}{2}\sqrt{2x - 3}\,dx$ **(2)**

    (c) $\int (4 - 3x)^{\frac{3}{2}}\,dx$ **(2)**

    (d) $\int \sqrt[3]{1 + 4x}\,dx$ **(2)**

5. Show that $\displaystyle\int_{2}^{6} \frac{1}{\sqrt{5x + 2}}\,dx = \frac{4}{5}(2\sqrt{2} - \sqrt{3})$ **(4)**

6. Find the value of $\displaystyle\int_{-1}^{0} \frac{1}{\sqrt[3]{1 - 7x}}\,dx$ **(4)**

Reverse chain rule

> **Guided**

1. Find

   (a) $\displaystyle\int \frac{3x-4}{3x^2-8x}\,dx$ **(3)**

   = ... $\ln(3x^2-8x)+c$

   (b) $\displaystyle\int 6x^2\,(7-4x^3)^6\,dx$ **(3)**

   = ... $(7-4x^3)^7+c$

2. Find

   (a) $\displaystyle\int x(6x^2-2)^5\,dx$ **(3)**

   (b) $\displaystyle\int \frac{3\sin 2\theta}{\cos 2\theta}\,d\theta$ **(3)**

3. Find

   (a) $\displaystyle\int 3\cos^3 x \sin x\,dx$ **(3)**

   (b) $\displaystyle\int \tfrac{1}{3}x\cos(x^2)\,dx$ **(3)**

4. Find

   (a) $\displaystyle\int 2\sec^2\theta \tan^3\theta\,d\theta$ **(3)**

   (b) $\displaystyle\int 3\operatorname{cosec}^5 x \cot x\,dx$ **(3)**

5. Find

   (a) $\displaystyle\int 4x\sqrt{3x^2-2}\,dx$ **(3)**

   (b) $\displaystyle\int \frac{2x^2-4x}{\sqrt{2x^3-6x^2+5}}\,dx$ **(3)**

6. Find the exact value of $\displaystyle\int_0^4 \frac{x}{x^2+2}\,dx$ **(5)**   Simplify your answer fully.

7. Find the exact value of $\displaystyle\int_0^{\ln 3} \frac{e^{-x}}{1+e^{-x}}\,dx$ **(5)**   Try $\ln|1+e^{-x}|$, and take care with the signs.

8. The function $f(x)$ is defined by $f(x)=\dfrac{2x^2-1}{4x^3-6x+5},\ x\geqslant 1$ **(5)**

   Find the area enclosed by $y=f(x)$, the lines $x=1$, $x=3$ and the x-axis.
   Give your answer correct to 3 decimal places.

Integrating partial fractions

> **Guided**

1. (a) Express $\dfrac{5 - 8x}{(2 + x)(1 - 3x)}$ in partial fractions. **(3)**

$\dfrac{5 - 8x}{(2 + x)(1 - 3x)} = \dfrac{A}{2 + x} + \dfrac{B}{1 - 3x}$ so $5 - 8x = A(1 - 3x) + B(2 + x)$

When $x = -2$, $21 = 7A$ so $A = 3$

When $x = $ ....

So $\dfrac{5 - 8x}{(2 + x)(1 - 3x)} \equiv \dfrac{3}{2 + x} + \dfrac{\text{..........}}{1 - 3x}$

> You can use the cover-up rule to find A and B.

(b) Hence find the exact value of $\displaystyle\int_{-1}^{0} \dfrac{5 - 8x}{(2 + x)(1 - 3x)}\,dx$,

giving your answer in the form $p\ln q$, where p and q are constants to be found. **(4)**

> Both of these are logarithm integrals. Use the rules of logarithms to simplify your answer.

$\displaystyle\int_{-1}^{0} \dfrac{5 - 8x}{(2 + x)(1 - 3x)}\,dx = \int_{-1}^{0} \left(\dfrac{3}{2 + x} + \dfrac{\text{..........}}{1 - 3x}\right) dx$

....

....

....

2. (a) Express $\dfrac{2x + 9}{4x^2 - 9}$ in the form $\dfrac{A}{2x + 3} + \dfrac{B}{2x - 3}$ where A and B are integers. **(3)**

....

....

(b) Show that $\dfrac{12x^3 - 31x - 18}{4x^2 - 9}$ can be written as $3x - \dfrac{2(2x + 9)}{4x^2 - 9}$ **(3)**

....

....

(c) Hence find $\displaystyle\int_{0}^{1} \dfrac{12x^3 - 31x - 18}{4x^2 - 9}\,dx$, giving your answer in the form $p + \ln q$,

where p and q are constants to be found. **(5)**

....

....

....

....

....

....

Identities in integration

Guided 1. Find the exact value of $\int_{\frac{\pi}{6}}^{\frac{\pi}{2}} \cos^2\left(\frac{x}{2}\right) dx$ **(5)**

$\cos 2A = 2\cos^2 A - 1$

When $A = \dfrac{x}{2}$, $\cos x = 2\cos^2\left(\dfrac{x}{2}\right) - 1$

So $\cos^2\left(\dfrac{x}{2}\right) = \dfrac{1}{2}(\cos x + 1)$

$\int_{\frac{\pi}{6}}^{\frac{\pi}{2}} \cos^2\left(\frac{x}{2}\right) dx = \int_{\frac{\pi}{6}}^{\frac{\pi}{2}} \frac{1}{2}(\cos x + 1) dx = \frac{1}{2}\left[\ldots\ldots + x\right]_{\frac{\pi}{6}}^{\frac{\pi}{2}}$

$\qquad = \frac{1}{2}\left[\left(\ldots\ldots + \frac{\pi}{2}\right) - \left(\ldots\ldots + \frac{\pi}{6}\right)\right] = \ldots\ldots$

> Leave your answer as an expression involving π.

2. Find the value of $\int_{0}^{\frac{\pi}{8}} \sin 2\theta \cos 2\theta \, d\theta$ **(5)**

> Use $\sin 2A = 2\sin A \cos A$ with $A = 2\theta$.

....

....

....

3. (a) By writing $\cos 5x$ as $\cos(3x + 2x)$ and $\cos x$ as $\cos(3x - 2x)$, show that $\cos 5x + \cos x = 2\cos 3x \cos 2x$ **(4)**

> Use the addition formulae.

....

....

(b) Hence find the value of $\int_{0}^{\frac{\pi}{6}} (\cos 3x \cos 2x) \, dx$ **(2)**

....

....

4. Find the exact value of $\int_{-\frac{\pi}{8}}^{\frac{\pi}{8}} \sin^2(2x) \, dx$ **(5)**

....

....

....

....

....

....

5. Find the exact value of $\int_{\frac{\pi}{6}}^{\frac{\pi}{3}} \cot^2(2x) \, dx$ **(5)**

....

....

....

....

....

....

Integration by substitution

Guided

1. Use the substitution $u = \sin 2x$ to find the exact value of $\int_{\frac{\pi}{12}}^{\frac{\pi}{4}} \sin^3 2x \cos 2x \, dx$ **(5)**

$u = \sin 2x$

$\dfrac{du}{dx} = 2\cos 2x \qquad$ so $\qquad \dfrac{du}{2} = \cos 2x \, dx$

When $x = \dfrac{\pi}{12}$, $u = \sin\dfrac{\pi}{6} = \ldots\ldots$ When $x = \dfrac{\pi}{4}$, $u = \sin\dfrac{\pi}{2} = \ldots\ldots$

$\int_{\frac{\pi}{12}}^{\frac{\pi}{4}} \sin^3 2x \cos 2x \, dx = \int \ldots\ldots\ldots\ldots\ldots$

$\qquad\qquad\qquad = \Big[\ldots\ldots\ldots\ldots\ldots\ldots\Big]$

$\qquad\qquad\qquad = \ldots\ldots\ldots\ldots = \ldots\ldots\ldots\ldots$

> Remember to transform the limits when using integration by substitution.

2. Use the substitution $x = 3\sin\theta$ to find the exact value of $\int_{\frac{3}{2}}^{\frac{3\sqrt{3}}{2}} \dfrac{1}{x^2\sqrt{9 - x^2}} \, dx$ **(5)**

....

....

....

....

....

....

3. Use the substitution $t = e^x$ to find the exact value of $\int_0^1 \dfrac{e^x}{(1 + e^x)^2} \, dx$ **(5)**

....

....

....

....

....

4. Use the substitution $x = 2\sin\theta$ to find the exact value of $\int_0^2 \sqrt{4 - x^2} \, dx$ **(7)**

> You will also need to use $\cos 2A = 2\cos^2 A - 1$

....

....

....

....

....

....

....

Integration by parts

> **Guided**

1. Use integration by parts to find $\int x\,e^{-x}\,dx$ **(4)** $\boxed{\int u\dfrac{dv}{dx}\,dx = uv - \int v\dfrac{dv}{dx}\,dx}$

$$u = x \qquad\qquad \frac{dv}{dx} = e^{-x}$$
$$\frac{du}{dx} = 1 \qquad\qquad v = -e^{-x}$$

$$\int x\,e^{-x}\,dx = x(-e^{-x}) - \int (-e^{-x})\,(1)\,dx$$

$$= -x\,e^{-x}\ldots\ldots\ldots\ldots\ldots = \ldots\ldots\ldots\ldots\ldots = \ldots\ldots\ldots\ldots\ldots$$

2. Use integration by parts to show that $\boxed{\text{Always set } \ln x \text{ as } u.}$

$$\int x^3 \ln x\,dx = \frac{x^4}{16}(4\ln x - 1) + c$$ **(4)**

....

....

....

....

....

....

3. (a) Use integration by parts to find $\int 2x \sin x\,dx$ **(3)**

....

....

....

(b) Use integration by parts and your answer to part (a) to find $\int x^2 \cos x\,dx$ **(4)**

....

....

....

....

4. Use integration by parts to find the exact value of $\int_1^2 x^2\,e^{2x}\,dx$ **(8)** $\boxed{\text{You will need to use integration by parts twice.}}$

....

....

....

....

....

....

....

....

Areas and parametric curves

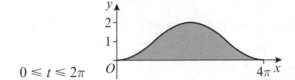

Guided

1. The diagram shows the curve with parametric equations

$$x = 2t - \sin t, \qquad y = 1 - \cos t, \qquad 0 \le t \le 2\pi$$

Find the area bounded by the curve and the x-axis. **(5)**

> Use $\cos 2A = 2\cos^2 A - 1$ to express $\cos^2 t$ in terms of $\cos 2t$. Tidy up then integrate.

$$x = 2t - 2\sin t, \qquad \frac{dx}{dt} = 2 - \cos t, \qquad y = 1 - \cos t$$

When $y = 0$, $\cos t = 1$, so $t = 0$ or 2π

$$\text{Area} = \int_{x=0}^{x=4\pi} y \, dx = \int_{t=0}^{t=2\pi} y \frac{dx}{dt} dt = \int_0^{2\pi} (1 - \cos t)(2 - \cos t) \, dt$$

$$= \int_0^{2\pi} (2 - 3\cos t + \cos^2 t) \, dt = \int_0^{2\pi} (2 - 3\cos t + \tfrac{1}{2} + \tfrac{1}{2}\cos 2t) \, dt$$

$$= \dots\dots\dots\dots\dots\dots\dots\dots = \dots\dots\dots\dots\dots\dots\dots\dots = \dots\dots\dots\dots\dots\dots\dots\dots$$

2.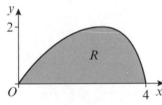

The diagram shows a curve with parametric equations

$$x = 4\cos t, \qquad y = 2\sin 2t, \qquad 0 \le t \le \frac{\pi}{2}$$

Find the area of the region marked R. **(8)**

> Write the integral in the form $k \int \cos t (\sin t)^2 \, dt$ then use the substitution $u = \sin t$.

....

....

....

....

....

....

3. The diagram shows a curve with parametric equations

$$x = t^3, \qquad y = e^t, \qquad 0 \le t \le 1$$

Find the area of the region marked R.

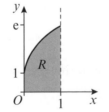

> You will need to use integration by parts twice.

(8)

....

....

....

....

....

....

....

Volumes of revolution 1

> **Guided**

1. The diagram shows the curve with equation $y = x^3 - 2x^2$

   The region, R, bounded by the curve, the x-axis, and the lines $x = 0$ and $x = 2$ is rotated 360° about the x-axis.

   Find the volume of the solid generated.   **(5)**

   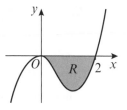

   $$V = \pi \int_0^2 y^2 \, dx = \pi \int_0^2 (x^3 - 2x^2)^2 \, dx = \pi \int_0^2 (x^6 - 4x^5 + 4x^4) \, dx$$

   $$= \pi \left[\frac{x^7}{7} - \ldots\ldots\ldots + \ldots\ldots\ldots \right]_0^2 = \ldots\ldots\ldots\ldots\ldots\ldots = \ldots\ldots\ldots\ldots$$

2. The diagram shows a curve with equation

   $$y = \sqrt{\frac{3x}{4x^2 - 1}}, \qquad x \geq 1$$

   The region, R, bounded by the curve, the x-axis, and the lines $x = 1$ and $x = 5$ is rotated 360° about the x-axis.

   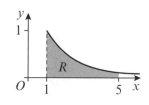

   Find the exact value of the volume of the solid generated, giving your answer in the form $p \ln q$ where p and q are constants.   **(5)**

   ....

   ....

   ....

3. A curve has equation $y = 3 \cos 2x$.

   The region, R, bounded by the curve, the x-axis, and the lines $x = \frac{\pi}{4}$ and $x = \frac{3\pi}{4}$ is rotated 2π radians about the x-axis.

   Find the volume of the solid generated.   **(6)**

   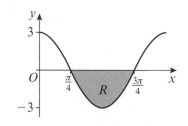

   ....

   ....

   You will need to use $\cos 2A = 2\cos^2 A - 1$

   ....

   ....

   ....

   ....

4. The region bounded by the curve $y = 3 - 2\cos x$, the x-axis and the lines $x = 0$ and $x = \pi$ is rotated 2π radians about the x-axis. Find the volume of the solid generated.   **(6)**

   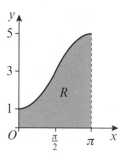

   ....

   ....

   ....

   ....

   ....

Volumes of revolution 2

Guided

1. The diagram shows the graph of a curve with parametric equations

$$x = t^2 - 1, \qquad y = t^3, \qquad t \geq 0$$

The region, R, bounded by the curve, the x-axis and the lines $x = 0$ and $x = 3$ is rotated 360° about the x-axis. Find the volume of the solid generated. **(5)**

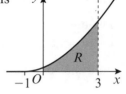

When $x = 0$, $t = 1$ When $x = 3$, $t = 2$

Volume $= \pi \displaystyle\int_0^3 y^2 \, dx = \pi \int_1^2 y^2 \frac{dx}{dt} \, dt = \pi \int_1^2 t^6 \, 2t \, dt = 2\pi \int_1^2 t^7 \, dt$

$= \ldots\ldots\ldots\ldots = \ldots\ldots\ldots\ldots = \ldots\ldots\ldots\ldots$

2. The curve C has parametric equations

$$x = t^3, \qquad y = 2t^2 + 1, \qquad t \geq 0$$

The line L intersects C at the point P, where $t = 1$.

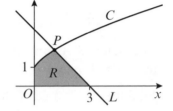

(a) Find the coordinates of P. **(1)**

....

....

(b) The finite region R is bounded by the curve C, the line L and the x- and y-axes. The region is rotated through 360° about the x-axis. Find the volume of the solid of revolution formed. **(7)**

> The volume of a cone with base radius r and height h is $V = \frac{1}{3}\pi r^2 h$.

....

....

....

....

3. The diagram shows the graph of a curve with parametric equations

$$x = t^2 + 2, \qquad y = \cos t, \qquad 0 \leq t \leq \frac{\pi}{2}$$

The region, R, bounded by the curve, the x-axis and the line $x = 2$ is rotated 2π radians about the x-axis. Find the volume of the solid generated. **(8)**

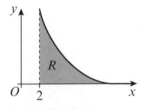

....

....

....

> You will need to use a double-angle identity and integration by parts.

....

....

....

....

The trapezium rule

> **Guided**

1. (a) Given that $y = \dfrac{x}{x^2 + 2}$ complete the table.

x	0	0.8	1.6	2.4	3.2	4
y	0	0.30303	0.35088			0.22222

(2)

(b) Use the trapezium rule with all the values of y in the completed table to obtain an estimate for $\displaystyle\int_0^4 \dfrac{x}{x^2 + 2}\,dx$

Show all the steps of your working and give your answer to 4 decimal places. **(3)**

$$\int_0^4 \frac{x}{x^2 + 2}\,dx \approx \frac{0.8}{2}[0 + 0.22222 + 2(0.30303 + \ldots\ldots\ldots + \ldots\ldots\ldots + \ldots\ldots\ldots)]$$

$$= \ldots\ldots\ldots\ldots\ldots \text{ (4 d.p.)}$$

(c) Integrate $\displaystyle\int_0^4 \dfrac{x}{x^2 + 2}\,dx$ giving your answer in the form $\ln k$ where k is an integer. **(5)**

....

....

(d) Calculate the % error in the estimate you obtained in part (b). **(2)**

....

....

2. The diagram shows a sketch of the curve
$$y = (2x - 1)\ln x, \qquad x > 0$$

(a) Complete this table of values.

x	1	1.5	2	2.5	3
y	0	0.81093			

(2)

(b) Use the trapezium rule with all the values of y in the completed table to obtain an estimate of the area under the curve between $x = 1$ and $x = 3$.
Show your working and give your answer to 4 decimal places. **(3)**

....

....

(c) Show that the exact value of $\displaystyle\int_1^3 (2x - 1)\ln x\,dx$ is $p\ln q + r$ where p, q and r are integers to be determined. **(7)**

....

....

....

....

....

....

....

....

Solving differential equations

Guided

1. (a) Show that $\dfrac{10}{(x-3)(3x+1)}$ can be written as $\dfrac{1}{x-3} - \dfrac{3}{3x+1}$ **(2)**

....

....

(b) Hence find $\displaystyle\int \dfrac{10}{(x-3)(3x+1)}\,dx$ where $x > 3$

> These are log integrals. You need not simplify your answer at this stage.

$$\int \frac{10}{(x-3)(3x+1)}\,dx = \int \left(\frac{1}{x-3} - \frac{3}{3x+1}\right)dx$$

$$= \text{..}$$ **(3)**

(c) Find the particular solution of the differential equation

> Separate the variables and integrate.

$$(x-3)(3x+1)\frac{dy}{dx} = 10y, \qquad x > 3$$

for which $y = 4$ at $x = 5$. Give your answer in the form $y = f(x)$.   **(6)**

....

....

....

....

....

....

2. The platform of a fairground ride oscillates vertically for the first 60 seconds of the ride.
   The height, h metres, and the time, t seconds, are connected by the differential equation

> You will need to use integration by parts.

$$\frac{dh}{dt} = \frac{t\cos\left(\frac{\pi}{8}t\right)}{40h}$$

At $t = 0$, the height of the platform is 3.5 metres.
Find the height of the platform after 36 seconds, giving your answer to the nearest centimetre.   **(9)**

....

....

....

....

....

....

....

....

Vectors

> **Guided**

1. The points P and Q have position vectors
$\mathbf{i} + 7\mathbf{j} + 5\mathbf{k}$ and $3\mathbf{i} + 2\mathbf{j} - 6\mathbf{k}$ respectively.

(a) Find $\left|\overrightarrow{OP}\right|$.   **(1)**

> Simplify your answer.

$\left|\overrightarrow{OP}\right| = \sqrt{1^2 + 7^2 + 5^2} = \sqrt{1 + \text{.................}} = \sqrt{\text{...........}} = \text{...........}$

(b) Find the vector $\overrightarrow{QP}$.   **(2)**

$\overrightarrow{QP} = \overrightarrow{QO} + \overrightarrow{OP}$

$= \overrightarrow{OP} - \overrightarrow{OQ}$

$= \mathbf{i} + 7\mathbf{j} + 5\mathbf{k} - (3\mathbf{i} + 2\mathbf{j} - 6\mathbf{k})$

$= \text{...........................}$

> The position vectors $\overrightarrow{OP}$ and $\overrightarrow{OQ}$ could be given in the form $\begin{pmatrix} 1 \\ 7 \\ 5 \end{pmatrix}$ and $\begin{pmatrix} 3 \\ 2 \\ -6 \end{pmatrix}$ respectively.

(c) Find the distance QP.   **(1)**

....

2. Find unit vectors in the direction of

(a) $8\mathbf{i} - 4\mathbf{j} - \mathbf{k}$ **(1)**   (b) $-6\mathbf{i} + 2\mathbf{j} + 3\mathbf{k}$ **(1)**   (c) $16\mathbf{i} - 8\mathbf{j} - 2\mathbf{k}$ **(1)**

....

....

> A unit vector has magnitude 1, so divide all coefficients by the magnitude of the given vector.

3. For each pair of position vectors $\overrightarrow{OP}$ and $\overrightarrow{OQ}$, find (i) vector $\overrightarrow{PQ}$ (ii) $\left|\overrightarrow{PQ}\right|$

(a) $\overrightarrow{OP} = 4\mathbf{i} + 3\mathbf{j}$
     $\overrightarrow{OQ} = \mathbf{i} - 2\mathbf{j} + \mathbf{k}$ **(3)**

(b) $\overrightarrow{OP} = 3\mathbf{i} + \mathbf{j} - \mathbf{k}$
     $\overrightarrow{OQ} = -3\mathbf{i} - \mathbf{j} + 8\mathbf{k}$ **(3)**

....

....

4. $\overrightarrow{OR} = \lambda\mathbf{i} - 3\mathbf{j} + 3\lambda\mathbf{k}$ where λ is a constant.
Given that $\left|\overrightarrow{OR}\right| = 7$, find the possible values of λ.   **(4)**

....

....

5. $\overrightarrow{OP} = \mathbf{i} + \mu\mathbf{j} - 4\mathbf{k}$, $\overrightarrow{OQ} = -3\mu\mathbf{i} + \mathbf{j} + \mathbf{k}$, μ is a constant
Given that $\left|\overrightarrow{QP}\right| = 5\sqrt{3}$, find the possible values of μ.   **(7)**

....

....

....

....

....

....

Vector equations of lines

Guided

1. With respect to a fixed origin, O, the point P has position vector $\mathbf{i} + 2\mathbf{j} - \mathbf{k}$ and the point Q has position vector $3\mathbf{i} - 3\mathbf{j} + 4\mathbf{k}$. The line L passes through P and Q.

   (a) Find $\overrightarrow{PQ}$. **(2)**

   $\overrightarrow{PQ} = \overrightarrow{PO} + \overrightarrow{OQ} = \overrightarrow{OQ} - \overrightarrow{OP} = 3\underset{\sim}{i} - 3\underset{\sim}{j} + 4\underset{\sim}{k} - (\underset{\sim}{i} + 2\underset{\sim}{j} - \underset{\sim}{k}) = $ ....

   (b) Find a vector equation for L. **(2)**

   Line L has equation $\underset{\sim}{r} = \underset{\sim}{i} + 2\underset{\sim}{j} - \underset{\sim}{k} + \lambda(\text{......................})$

2. For each pair of position vectors $\overrightarrow{OP}$ and $\overrightarrow{OQ}$, find the vector equation of the line passing through P and Q.

   (a) $\overrightarrow{OP} = 2\mathbf{i} - 3\mathbf{j} + \mathbf{k}$

   $\overrightarrow{OQ} = 2\mathbf{i} + \mathbf{j} + \mathbf{k}$ **(4)**

   (b) $\overrightarrow{OP} = \mathbf{i} + 5\mathbf{j} - 6\mathbf{k}$

   $\overrightarrow{OQ} = 4\mathbf{i} - 2\mathbf{j} - \mathbf{k}$ **(4)**

   ....

   ....

   ....

   ....

3. A line, L, has vector equation

   $$\mathbf{r} = \begin{pmatrix} 10 \\ -3 \\ 4 \end{pmatrix} + \lambda \begin{pmatrix} -4 \\ -2 \\ 3 \end{pmatrix}$$

   Point $P(a, 5, -8)$ and point $Q(2, b, 10)$ both lie on L.
   Find the values of the constants a and b. **(3)**

   ....

   ....

   ....

   ....

4. The point P with coordinates $(-3, 8, a)$ lies on the line L with vector equation

   $$\mathbf{r} = (7\mathbf{i} + b\mathbf{j} - 4\mathbf{k}) + \lambda(2\mathbf{i} - 3\mathbf{j} + 5\mathbf{k})$$

   (a) Find the values of the constants a and b. **(3)**

   ....

   ....

   (b) Point Q lies on L where $\lambda = -3$. Find the coordinates of Q. **(1)**

   ....

   ....

   ....

Intersecting lines

> **Guided**

1. The line L_1 has equation $\mathbf{r} = \begin{pmatrix} 4 \\ -5 \\ 3 \end{pmatrix} + \lambda \begin{pmatrix} -1 \\ 3 \\ 1 \end{pmatrix}$

and the line L_2 has equation $\mathbf{r} = \begin{pmatrix} 7 \\ -8 \\ 6 \end{pmatrix} + \mu \begin{pmatrix} 2 \\ -3 \\ 1 \end{pmatrix}$

λ and μ are scalar parameters.
Show that the lines L_1 and L_2 intersect and find the position vector of their point of intersection. **(6)**

$$\begin{pmatrix} 4 \\ -5 \\ 3 \end{pmatrix} + \lambda \begin{pmatrix} -1 \\ 3 \\ 1 \end{pmatrix} = \begin{pmatrix} 7 \\ -8 \\ 6 \end{pmatrix} + \mu \begin{pmatrix} 2 \\ -3 \\ 1 \end{pmatrix}$$

$4 - \lambda = 7 + 2\mu$ ①

.... = ...   ②

$3 + \lambda = 6 + \mu$ ③

> You need to check that your values of λ and μ satisfy all three equations simultaneously.

① + ③: ...., so $\mu = $ ...., hence $\lambda = $ ....

Substitute $\lambda = $ .... and $\mu = $ .... into ②:

LHS = ...   RHS = ...

Using L_1, position vector of point of intersection is $\begin{pmatrix} .... \\ .... \\ .... \end{pmatrix} + ... \begin{pmatrix} .... \\ .... \\ .... \end{pmatrix} = \begin{pmatrix} .... \\ .... \\ .... \end{pmatrix}$

2. The line L_1 has equation $\mathbf{r} = \begin{pmatrix} -2 \\ 3 \\ 2 \end{pmatrix} + \lambda \begin{pmatrix} 1 \\ -2 \\ 1 \end{pmatrix}$

and the line L_2 has equation $\mathbf{r} = \begin{pmatrix} 1 \\ 5 \\ -4 \end{pmatrix} + \mu \begin{pmatrix} -2 \\ 4 \\ -2 \end{pmatrix}$

(a) Explain why lines L_1 and L_2 are parallel. **(2)**

....

....

(b) A is the point on L_1 with $\lambda = 3$ and B is the point on L_2 with $\mu = -5$.
C is the point with position vector $-19\mathbf{i} + 21\mathbf{j} + 3\mathbf{k}$.
Show that A, B and C are collinear. **(6)**

> You need to find the vector equation of the line passing through A and B. Then check that C lies on the line.

....

....

....

....

....

....

....

Scalar product

> **Guided**

1. Relative to a fixed origin, points P and Q have position vectors

$$\overrightarrow{OP} = -3\mathbf{i} + 5\mathbf{j} - 2\mathbf{k} \quad \text{and} \quad \overrightarrow{OQ} = 2\mathbf{i} + 4\mathbf{j} + \mathbf{k}$$

Find the size of angle POQ, giving your answer to 1 decimal place. **(3)**

$\overrightarrow{OP} \cdot \overrightarrow{OQ} = (-3)(2) + (5)(4) + (-2)(1) = \dots\dots\dots\dots$

$\left|\overrightarrow{OP}\right| = \sqrt{(-3)^2 + 5^2 + (-2)^2} = \dots\dots\dots\dots$

$\left|\overrightarrow{OQ}\right| = \sqrt{\dots\dots\dots\dots\dots\dots\dots\dots} = \dots\dots\dots\dots$

$\cos POQ = \dfrac{\overrightarrow{OP} \cdot \overrightarrow{OQ}}{\left|\overrightarrow{OP}\right|\left|\overrightarrow{OQ}\right|} = \dfrac{\dots\dots\dots\dots}{\dots\dots\dots\dots}$ so angle $POQ = \dots\dots\dots\dots$

> Use brackets when calculating a scalar product to avoid making a mistake with negative signs.

2. With respect to a fixed origin, points A, B and C have position vectors $3\mathbf{i} + \mathbf{j} - 6\mathbf{k}$, $5\mathbf{i} - 2\mathbf{j}$ and $8\mathbf{i} - 4\mathbf{j} - 6\mathbf{k}$ respectively. Find the size of angle ACB, giving your answer to 1 decimal place. **(4)**

> Find $\overrightarrow{CA}$ and $\overrightarrow{CB}$, then work out $\overrightarrow{CA} \cdot \overrightarrow{CB}$, $\left|\overrightarrow{CA}\right|$ and $\left|\overrightarrow{CB}\right|$.

....

....

....

....

....

....

3. Points D and E have coordinates $(4, -1, 5)$ and $(-2, 2, 3)$ respectively. The line L passes though D and has equation

$$\mathbf{r} = \begin{pmatrix} 4 \\ -1 \\ 5 \end{pmatrix} + \lambda \begin{pmatrix} 2 \\ 8 \\ -1 \end{pmatrix}$$

(a) Find the vector $\overrightarrow{DE}$. **(2)**

....

....

....

(b) Find the size of the acute angle between DE and the line L, giving your answer to 1 decimal place. **(3)**

....

....

....

....

....

Perpendicular vectors

> **Guided**

1. Points A, B and C have position vectors
$$\overrightarrow{OA} = 5\mathbf{i} + 3\mathbf{j} + 2\mathbf{k}, \qquad \overrightarrow{OB} = 2\mathbf{i} - \mathbf{j} + 3\mathbf{k}, \qquad \overrightarrow{OC} = 7\mathbf{i} - 3\mathbf{j} + 10\mathbf{k}$$
Show that angle ABC is a right angle. **(4)**

$\overrightarrow{BA} = \overrightarrow{BO} + \overrightarrow{OA} = -2\underset{\sim}{\mathrm{i}} + \underset{\sim}{\mathrm{j}} - 3\underset{\sim}{\mathrm{k}} + 5\underset{\sim}{\mathrm{i}} + 3\underset{\sim}{\mathrm{j}} + 2\underset{\sim}{\mathrm{k}} = 3\underset{\sim}{\mathrm{i}} + \dots\dots\dots\dots$

$\overrightarrow{BC} = \dots\dots\dots\dots = \dots\dots\dots\dots\dots\dots\dots\dots\dots = \dots\dots\dots\dots\dots\dots$

$\overrightarrow{BA} \cdot \overrightarrow{BC} = (3)(\dots\dots) + (\dots\dots)(\dots\dots) + (\dots\dots)(\dots\dots) = \dots\dots\dots\dots$ so $\angle ABC = 90°$

2. With respect to a fixed origin, O, the lines L_1 and L_2 have equations
$$L_1: \mathbf{r} = \begin{pmatrix} 5 \\ -1 \\ -2 \end{pmatrix} + \lambda \begin{pmatrix} 2 \\ -3 \\ -6 \end{pmatrix} \qquad L_2: \mathbf{r} = \begin{pmatrix} 5 \\ -1 \\ -2 \end{pmatrix} + \mu \begin{pmatrix} 3 \\ -6 \\ -2 \end{pmatrix}$$

where λ and μ are scalar parameters.

The lines intersect at A and the acute angle between them is θ.

(a) Write down the coordinates of A. **(1)**

....

(b) Find the value of θ, giving your answer to 1 decimal place. **(3)**

....

....

....

The point B lies on L_1 where $\lambda = 3$.

(c) Find the coordinates of B. **(1)**

....

....

(d) Find the vector $\overrightarrow{AB}$. **(2)**

....

....

....

(e) Hence find $\left|\overrightarrow{AB}\right|$. **(2)**

....

The point C lies on L_2 such that $\overrightarrow{BC}$ is perpendicular to L_1.

(f) Find the length AC, giving your answer to 1 decimal place. **(3)**

....

....

....

....

Solving area problems

Guided 1. Three points A, B and C are such that

$$\overrightarrow{BA} = 2\mathbf{i} + 4\mathbf{j} - 5\mathbf{k} \qquad \text{and} \qquad \overrightarrow{BC} = 4\mathbf{i} + 2\mathbf{j} - 6\mathbf{k}$$

Find the area of triangle ABC to 2 decimal places. **(4)**

$\left|\overrightarrow{BA}\right| = \sqrt{(2)^2 + (4)^2 + (-5)^2} = \ldots\ldots\ldots\ldots \qquad \left|\overrightarrow{BC}\right| = \sqrt{\rule{4cm}{0.4pt}} = \ldots\ldots\ldots\ldots$

$\cos ABC = \dfrac{\overrightarrow{BA} \cdot \overrightarrow{BC}}{\left|\overrightarrow{BA}\right|\left|\overrightarrow{BC}\right|} = \dfrac{(2)(4) + (\ldots\ldots)(\ldots\ldots) + (\ldots\ldots)(\ldots\ldots)}{(\ldots\ldots)(\ldots\ldots)} = \ldots\ldots\ldots$, so $\angle ABC = \ldots\ldots\ldots$

Area of $\triangle ABC = \frac{1}{2} \times \left|\overrightarrow{BA}\right| \times \left|\overrightarrow{BC}\right| \times \sin ABC = \frac{1}{2} \times \ldots\ldots\ldots \times \ldots\ldots\ldots \times \sin\ldots\ldots = \ldots\ldots\ldots$

2. The points A and B have coordinates $(3, -4, 2)$ and $(1, 0, -2)$ respectively.

The line L passes through A and has equation $\mathbf{r} = \begin{pmatrix} 3 \\ -4 \\ 2 \end{pmatrix} + \lambda \begin{pmatrix} -3 \\ 5 \\ 2 \end{pmatrix}$

(a) Find the vector $\overrightarrow{AB}$. **(2)**

....

....

(b) Find the acute angle between AB and the line L, to 1 decimal place. **(3)**

....

....

....

(c) Point C lies on L and is such that angle ABC is a right angle. Find the coordinates of the point C. **(4)**

Use the fact that $\overrightarrow{AB} \cdot \overrightarrow{BC} = 0.$

....

....

....

....

....

....

(d) Find the area of triangle ABC, to 1 decimal place. **(3)**

....

....

....

....

(e) Given that $ABCD$ is a rectangle, find the coordinates of point D. **(2)**

....

....

You are the examiner!

Checking your work is a key skill for A2 Maths. Have a look at pages 53 and 54 of the *Revision Guide*, then practise with these questions. There are full worked solutions on page 67.

1. Express $\dfrac{x - 5}{(x + 1)^2(x - 1)}$ in partial fractions. **(4)**

> There is a repeated factor so you will need to substitute values of x and/or compare coefficients.

....

....

....

....

....

....

....

2. Find the exact value of $\displaystyle\int_{\frac{\pi}{4}}^{\frac{\pi}{2}} 5x \sin 2x \, dx$ **(6)**

> 'Find the exact value' means leave your answer as an expression involving π.

....

....

....

....

....

....

....

....

3. A curve, C, has equation
$$y^3 - xy^2 - x^3 = 3$$

(a) Find $\dfrac{dy}{dx}$ in terms of x and y. **(4)**

> Differentiate every term implicitly, including the constant on the right-hand side.

....

....

....

....

....

....

....

(b) Hence find the gradient of the normal to C at the point $(1, 2)$. **(2)**

....

....

You are the examiner!

> Checking your work is a key skill for A2 Maths. Have a look at pages 53 and 54 of the *Revision Guide*, then practise with these questions. There are full worked solutions on page 67.

4. Relative to a fixed origin, O, points P and Q have position vectors $2\mathbf{i} - 5\mathbf{j} - \mathbf{k}$ and $4\mathbf{i} + \mathbf{j} - 2\mathbf{k}$ respectively. The line L_1 passes through P and Q.

   (a) Find a vector equation for the line L_1. **(3)**

   ....

   ....

   ....

   ....

   ....

   (b) The line L_2 passes through Q and has equation $\mathbf{r} = \begin{pmatrix} 4 \\ 1 \\ -2 \end{pmatrix} + \mu \begin{pmatrix} 0 \\ 1 \\ -3 \end{pmatrix}$

   Find the acute angle between L_1 and L_2, giving your answer to 1 decimal place. **(3)**

   ....

   ....

   | Remember to use the direction vectors of L_1 and L_2 in the scalar product. |

   ....

   ....

   ....

   ....

   ....

5. Use the substitution $u^2 = 2x - 1$ to show that

   $$\int x\sqrt{2x - 1}\, \mathrm{d}x = \tfrac{1}{15}(2x - 1)^{\frac{3}{2}}(3x + 1) + c \text{ where } c \text{ is a constant.} \quad \textbf{(5)}$$

   ....

   | Use implicit differentiation to find the relationship between $\mathrm{d}u$ and $\mathrm{d}x$. After integrating, give your answer in terms of the original variable and use algebra to show the required result. |

   ....

   ....

   ....

   ....

   ....

   ....

   ....

   ....

Calculators may be used in this practice paper.

Time: 1 hour 30 minutes

Total marks: 75

You may use the Mathematical Formulae and Statistical Tables booklet which is available from the Edexcel website.

---

1. Express as a single fraction in its simplest form

$$\frac{4}{x-1} - \frac{2x-6}{2x^2-5x+3}$$

(5)

---

2. The area, A mm^2, of a bacterial culture growing in a solution, t hours after midday, is given by

$$A = 18\,\mathrm{e}^{1.6t}, \qquad t \geq 0$$

(a) Write down the area of the culture at midday. **(1)**

(b) Find the time at which the area of the culture is twice its area at midday.
Give your answer to the nearest minute. **(5)**

---

3. The diagram shows the graph of $y = \mathrm{f}(x), \ x \in \mathbb{R}$

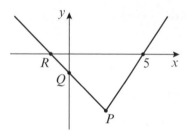

The graph consists of two line segments that meet at the point P.
The graph cuts the y-axis at the point Q and the x-axis at the points R and $(5, 0)$.
Sketch, on separate diagrams, the graphs of

(a) $y = |\mathrm{f}(x)|$ **(2)**

(b) $y = \mathrm{f}(-x)$ **(2)**

(c) Given that $\mathrm{f}(x) = |x - 2| - 3$, find the coordinates of the points P, Q and R. **(3)**

---

4. (i) $y = \mathrm{e}^{3x}(2x^2 + 1)^{\frac{3}{2}}$

Find the value of $\dfrac{\mathrm{d}y}{\mathrm{d}x}$ when $x = 0$. **(5)**

(ii) A curve, C, has equation $y = \dfrac{3x - 4}{x^2 + 1}$

Find the x-coordinates of the turning points on the curve, C. **(5)**

5. $f(x) = 3 \ln x - 2\sqrt{x}$

    (a) Show that $f(x) = 0$ has a root, $x = \alpha$, lying between 3 and 4. **(2)**

    (b) Show that $3 \ln x = 2\sqrt{x}$ can be arranged into the form $x = e^{\frac{2}{3}\sqrt{x}}$ **(1)**

    The iterative formula $x_{n+1} = e^{\frac{2}{3}\sqrt{x_n}}$ with $x_0 = 3$, is used to find an approximation for α.

    (c) Calculate the values of x_1, x_2 and x_3 to 4 decimal places. **(3)**

    (d) By choosing a suitable interval, show that $\alpha = 3.449$ correct to 3 decimal places. **(3)**

---

6. The curve, C, has equation $x = 6y \tan 2y$

    The point, P, has coordinates $\left(\sqrt{3}\pi, \frac{\pi}{6} \right)$

    (a) Verify that P lies on C. **(1)**

    (b) Find an equation of the tangent to C at P in the form $ay = x + b$, where the constants a and b are to be found in terms of π. **(7)**

---

7. The function f is defined by $f : x \mapsto x^2 - 4x + 3, \ x \geqslant 2$

    (a) Find the range of f. **(2)**

    (b) Find $f^{-1}(x)$. **(4)**

    The function g is defined by $g : x \mapsto |x - 4|, \ x \in \mathbb{R}$

    (c) Write down an expression for $gf(x)$. **(1)**

    (d) Solve $gf(x) = 4$ **(4)**

---

8. (a) Express $11 \sin \theta + 7 \cos \theta$ in the form $R \cos(\theta - \alpha)$ where R and α are constants and $R > 0$, $0 < \alpha < 90°$, giving the exact value of R and the value of α to 1 decimal place. **(3)**

    (b) Hence, or otherwise, solve the equation $11 \sin \theta + 7 \cos \theta = 12$, for $0 \leqslant \theta \leqslant 360°$, giving your answers to 1 decimal place. **(5)**

---

9. (a) Without using a calculator, find the exact value of $(\sin 67.5° + \cos 67.5°)^2$
    You must show each stage of your working. **(4)**

    (b) Solve $2 \cos 4x = 1 - 4 \cos 2x$ for $0 \leqslant x \leqslant 2\pi$ **(7)**

Calculators may be used in this practice paper.
Time: 1 hour 30 minutes
Total marks: 75
You may use the Mathematical Formulae and Statistical Tables booklet which is available from the Edexcel website.

1. The diagram shows a sketch of part of the curve with equation $y = 4\cos\left(\frac{x}{2}\right)$, $0 \leqslant x \leqslant \pi$

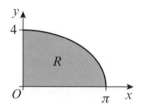

The finite region R, shown in the diagram, is bounded by the curve and the x- and y-axes.

The table shows corresponding values of x and y for the equation $y = 4\cos\left(\frac{x}{2}\right)$

x	0	$\frac{\pi}{4}$	$\frac{\pi}{2}$	$\frac{3\pi}{4}$	π
y	4	3.69552	2.82843		0

(a) Complete the table by giving the missing value of y to 5 decimal places. **(1)**

(b) Use the trapezium rule with all the values of y in the completed table to obtain an estimate for the area of R, giving your answer to 4 decimal places. **(3)**

(c) Find the exact value of the area, R, by evaluating $\int_0^\pi 4\cos\left(\frac{x}{2}\right) dx$. **(2)**

(d) Calculate the % error in the estimate you obtained in part (b). **(2)**

---

2. Use the substitution $u = 4 - x^3$ to find the exact value of $\int_0^1 \frac{x^5}{4 - x^3} dx$ **(7)**

---

3. (a) Given that $f(x) = \frac{8x + 2}{(1 - 3x)(2 + x)}$, express $f(x)$ in the form $\frac{A}{1 - 3x} + \frac{B}{2 + x}$ where A and B are integers. **(3)**

(b) Find the first three terms in the binomial expansion of $f(x)$ in the form $a + bx + cx^2$ where a, b and c are rational numbers. **(7)**

---

4. (a) Use integration by parts to find $\int t\,e^t\,dt$ **(3)**

(b) Hence find $\int_0^1 (t - t^2)e^t\,dt$ **(6)**

5. The diagram shows a sketch of part of the curve with equation $y = \dfrac{5}{\sqrt{1-2x}}$

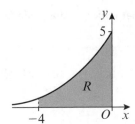

The finite region R, shown shaded in the diagram, is bounded by the curve, the x-axis and the lines with equations $x = -4$ and $x = 0$.

(a) Use integration to find the area of R. **(4)**

The region R is rotated $360°$ about the x-axis.

(b) Use integration to find the exact volume of the solid formed. **(5)**

6. A curve, C, has equation $3ye^{-x} = 4x + y^2$

(a) Find $\dfrac{dy}{dx}$ **(5)**

Point P on C has coordinates $(0, 3)$.

(b) Find the equation of the normal at P giving your answer in the form $ax + by + c = 0$ where a, b and c are integers. **(4)**

7. A differential equation is given as $\dfrac{dx}{dt} = -\dfrac{kt^2}{2}\, e^{\frac{1}{3}x}$ where k is a positive constant.

(a) Given that $x = 6$ when $t = 0$, solve the differential equation to show that

$$x = -3\ln\left(\frac{kt^3}{18} + e^{-2}\right)$$ **(6)**

(b) The population of a colony of insects is decreasing according to the model

$$\frac{dx}{dt} = -\frac{kt^2}{2}\, e^{\frac{1}{3}x}$$

where x, in thousands, is the number of insects in the colony after time t minutes. Initially there were 6000 insects.
Given that $k = 0.009$, find
(i) the population of the colony after 8 minutes, giving your answer to the nearest hundred **(2)**
(ii) the time after which there will be no insects left in the colony, giving your answer to the nearest minute. **(2)**

8. With respect to a fixed origin, O, the lines L_1 and L_2 are given by the equations

$$L_1: \mathbf{r} = \begin{pmatrix} 7 \\ 6 \\ -5 \end{pmatrix} + \lambda \begin{pmatrix} 5 \\ -1 \\ -3 \end{pmatrix} \qquad L_2: \mathbf{r} = \begin{pmatrix} -6 \\ 1 \\ 2 \end{pmatrix} + \mu \begin{pmatrix} 4 \\ 3 \\ -2 \end{pmatrix}$$

where λ and μ are scalar parameters.

(a) Given that L_1 and L_2 meet, find the position vector of their point of intersection, P. **(5)**

(b) Find, to the nearest $0.1°$, the acute angle between the lines L_1 and L_2. **(3)**

The point A lies on L_1 where $\lambda = 1$.

(c) Write down the position vector of the point A. **(1)**

(d) Find the shortest distance from A to the line L_2, giving your answer to 3 significant figures. **(4)**

Answers

Worked solutions have been provided for all Guided questions. These are marked with a ⟩**G**⟩.
Short answers have been provided for all other questions.

Core Mathematics 3

1 Algebraic fractions

⟩**G**⟩ 1. $\dfrac{2x^2 + 7x - 30}{x^2 - 36} = \dfrac{(2x - 5)(x + 6)}{(x + 6)(x - 6)} = \dfrac{2x - 5}{x - 6}$

2. $\dfrac{3x + 2}{x - 1}$

3. $\dfrac{2(x - 4)}{5x - 3}$

4. $\dfrac{2x + 1}{x(x - 1)}$

5. $\dfrac{3x - 5}{x - 5}$

6. $\dfrac{3(2x - 5)}{(3x + 2)(x - 3)}$

2 Algebraic division

⟩**G**⟩ 1.

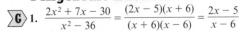

$$x^2 + 0x - 3 \,\overline{)\,2x^4 - 7x^3 - 10x^2 + 24x + 10}$$
$$\underline{2x^4 + 0x^3 - \ 6x^2}$$
$$-7x^3 - \ 4x^2 + 24x + 10$$
$$\underline{-7x^3 + \ 0x^2 + 21x}$$
$$-4x^2 + \ 3x + 10$$
$$\underline{-4x^2 + \ 0x + 12}$$
$$3x - \ 2$$

Answer: $2x^2 - 7x - 4 + \dfrac{3x - 2}{x^2 - 3}$

2. $a = 2, b = -5, c = 4, d = -11, e = 14$
3. $a = 5, b = 3, c = 10, d = 16, e = 12$

3 Functions

⟩**G**⟩ 1. (a) $\mathrm{gf}(x) = 2 + \dfrac{3}{x^2 - 2} = \dfrac{2(x^2 - 2) + 3}{x^2 - 2} = \dfrac{2x^2 - 1}{x^2 - 2}$

(b) $\frac{2}{7}$

(c) $x = \dfrac{1}{\sqrt{2}}$ and $x = -\dfrac{1}{\sqrt{2}}$

2. (a) $\mathrm{gf}(x) = \dfrac{2(1 - 5x)}{(1 - 5x) + 3} = \dfrac{2 - 10x}{4 - 5x}$

(b) 1.75

(c) $\mathrm{gg}(x) = \dfrac{4x}{5x + 9}$

(d) -1

(e) $x = \frac{1}{5}$ and $x = \frac{2}{5}$

4 Graphs and range

⟩**G**⟩ 1. (a) Range of f is $5 < \mathrm{f}(x) < 11$

(b) $\mathrm{ff}(6) = \mathrm{f}(-4) = -6$

2. (a) $0 < \mathrm{f}(x) < \frac{2}{3}$

(b) $-12 \leqslant \mathrm{g}(x) < 24$

3. (a) $\dfrac{5x + 14}{x^2 + 4x - 12} - \dfrac{3}{x - 2} = \dfrac{5x + 14}{(x + 6)(x - 2)} - \dfrac{3}{x - 2}$

$= \dfrac{5x + 14 - 3(x + 6)}{(x + 6)(x - 2)}$

$= \dfrac{2x - 4}{(x + 6)(x - 2)}$

$= \dfrac{2(x - 2)}{(x + 6)(x - 2)}$

$= \dfrac{2}{x + 6}$

(b) $\frac{1}{4} < \mathrm{f}(x) < 2$

5 Inverse functions

⟩**G**⟩ 1. (a) $y = \dfrac{2x}{3} - 4, 3y = 2x - 12, x = \frac{1}{2}(3y + 12)$

so the inverse function f^{-1} is $x \mapsto \frac{1}{2}(3x + 12)$

(b) Domain of f^{-1} is $-6 < x < 2$

2. (a) $\mathrm{g}^{-1}(x) = \dfrac{6x + 5}{x - 2}$

(b) Domain of g^{-1} is $2 < x < 10.5$

3. (a) $\mathrm{h}^{-1}(x) = \dfrac{4 - x}{3x - 1}$

(b) Domain of h^{-1} is $\frac{1}{3} < x < 4$

6 Inverse graphs

⟩**G**⟩ 1.

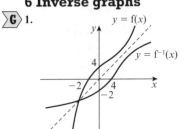

2.

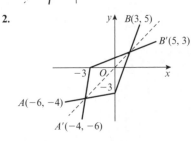

3. A has an inverse function.
   B does not have an inverse function.
   C does not have an inverse function.
   D has an inverse function.
   Those that do not have an inverse function are not one-to-one mappings – a line drawn horizontally cuts the graph more than once.

7 Modulus

⟩**G**⟩ 1. (a)

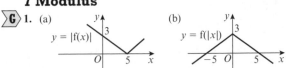

⟩**G**⟩ 2. (a) (i)

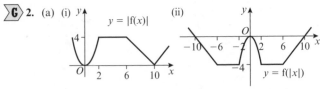

(b) (i)

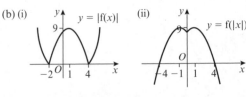

(c) (i)

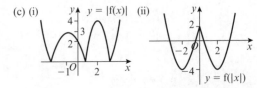

8 Transformations of graphs

1.

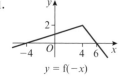

$y = f(-x)$

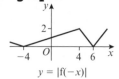

$y = |f(-x)|$

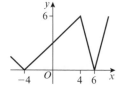

$y = 3|f(-x)|$

2. (a)

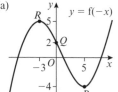

$y = f(-x)$ $y = f(-x) + 4$

(b)

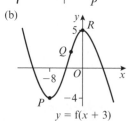

$y = f(x+3)$ $y = |f(x+3)|$

9 Modulus equations

1. (a)

$y = 7 - 2x$ $y = |7 - 2x|$

(b) $7 - 2x = x + 3$ $-(7 - 2x) = x + 3$
$x = \frac{4}{3}$ $x = 10$

2. $x = 4$ and $x = \frac{4}{9}$

3. (a)

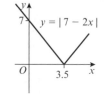

$y = -x$ $y = f(x) = |6 - 3x|$

(b) The graph of $y = -x$ does not intersect the graph of $y = f(x)$.

(c) $x = 1.5$ and $x = 3$

10 Sec, cosec and cot

1.

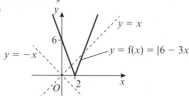

$y = \operatorname{cosec}\theta$

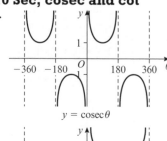

$y = 2\operatorname{cosec}\left(\frac{1}{2}\theta\right)$

2.

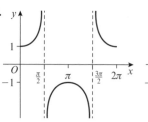

$y = \sec x$ $y = 2\sec\left(x - \frac{\pi}{2}\right)$

3.

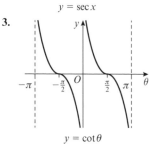

$y = \cot\theta$ $y = \cot\left(\theta + \frac{\pi}{2}\right) - 1$

11 Trig equations 1

1. $\operatorname{cosec} 2x = \frac{2}{\sqrt{3}}, \frac{1}{\sin 2x} = \frac{2}{\sqrt{3}}, \sin 2x = \frac{\sqrt{3}}{2}$
$-180° \leq x \leq 180°$ so $-360° \leq 2x \leq 360°$
$2x = -300°, -240°, 60°, 120°$
$x = -150°, -120°, 30°, 60°$

2. $\theta = \frac{\pi}{12}$ and $\theta = \frac{7\pi}{12}$

3. $x = 78.5°, 281.5°$

4. $\theta = 23.6°, 156.4°, 270.0°$

12 Using trig identities

1. $\operatorname{cosec}\theta - \sin\theta \equiv \frac{1}{\sin\theta} - \sin\theta \equiv \frac{1 - \sin^2\theta}{\sin\theta} \equiv \frac{\cos^2\theta}{\sin\theta}$
$\equiv \frac{\cos\theta}{\sin\theta} \times \cos\theta \equiv \cot\theta\cos\theta$

2. $\theta = 0, \frac{3\pi}{4}, \pi, \frac{7\pi}{4}, 2\pi$

3. $\sec^2\theta + \operatorname{cosec}^2\theta \equiv \frac{1}{\cos^2\theta} + \frac{1}{\sin^2\theta} \equiv \frac{\sin^2\theta + \cos^2\theta}{\cos^2\theta\sin^2\theta}$
$\equiv \frac{1}{\cos^2\theta\sin^2\theta} \equiv \sec^2\theta\operatorname{cosec}^2\theta$

4. $\theta = 22.5°, 38.0°, 112.5°, 128.0°$

5. $\theta = 18.4°, 135.0°, 198.4°, 315.0°$

13 Arcsin, arccos and arctan

1. (a) $\frac{\pi}{6}$ (b) $\frac{5\pi}{6}$ (c) $\frac{-\pi}{4}$

2. (a) $f\left(\frac{\sqrt{3}}{2}\right) = \arccos\left(\frac{\sqrt{3}}{2}\right) + \frac{\pi}{4} = \frac{\pi}{6} + \frac{\pi}{4} = \frac{5\pi}{12}$

(b) $x = \frac{1}{2}$

(c) Let $y = \arccos x + \frac{\pi}{4}$, $y - \frac{\pi}{4} = \arccos x$, $\cos\left(y - \frac{\pi}{4}\right) = x$
So $f^{-1}(x) = \cos\left(x - \frac{\pi}{4}\right)$
Domain of f^{-1} = range of f, so $\frac{\pi}{4} \leq x \leq \frac{5\pi}{4}$

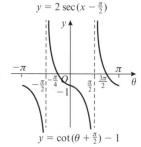

$y = f^{-1}(x)$

3. $g^{-1}(x) = \sin\left(\frac{x}{2} - \frac{\pi}{4}\right)$
Domain is $\frac{\pi}{2} \leq x \leq \frac{3\pi}{2}$

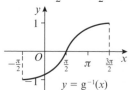

$y = g^{-1}(x)$

14 Addition formulae

G **1.** $\tan 105° = \tan(60° + 45°) = \dfrac{\tan 60° + \tan 45°}{1 - \tan 60° \tan 45°}$

$$= \dfrac{\sqrt{3} + 1}{1 - (\sqrt{3})(1)} = \dfrac{(\sqrt{3} + 1)(1 + \sqrt{3})}{(1 - \sqrt{3})(1 + \sqrt{3})}$$

$$= \dfrac{4 + 2\sqrt{3}}{1 - 3} = -2 - \sqrt{3}$$

2. (a) $\theta = 30°, 210°$
(b) $\theta = 26.6°, 108.4°, 206.6°, 288.4°$

3. (a) $\dfrac{-29}{35}$

(b) $\dfrac{-4\sqrt{24}}{35}$ or $\dfrac{-8\sqrt{6}}{35}$

15 Double angle formulae

G **1.**
$$1 = 2\sin\theta + 4(1 - 2\sin^2\theta)$$
$$8\sin^2\theta - 2\sin\theta - 3 = 0$$
$$(4\sin\theta - 3)(2\sin\theta + 1) = 0$$
$$\sin\theta = \tfrac{3}{4} \text{ or } \sin\theta = -\tfrac{1}{2}$$
$$\theta = 48.6°, 131.4°, \theta = -30°, -150°$$

2. (a) $\dfrac{-4\sqrt{5}}{9}$ (b) $\dfrac{1}{9}$ (c) $-4\sqrt{5}$

3. $\dfrac{\sin 3A}{\sin A} + \dfrac{\cos 3A}{\cos A} = \dfrac{\sin 3A \cos A + \cos 3A \sin A}{\sin A \cos A}$

$$= \dfrac{\sin 4A}{\tfrac{1}{2}\sin 2A} = \dfrac{2\sin 2A \cos 2A}{\tfrac{1}{2}\sin 2A} = 4\cos 2A$$

4. $x = 129.6°, 230.4°$ (1 d.p.)

16 $a\cos\theta \pm b\sin\theta$

G **1.** (a) $4\cos\theta + 2\sin\theta \equiv R(\cos\theta\cos\alpha + \sin\theta\sin\alpha)$
$R\cos\alpha = 4, R\sin\alpha = 2, \tan\alpha = \tfrac{1}{2}, \alpha = 26.6°$
$R = \sqrt{4^2 + 2^2} = \sqrt{20}$
So $4\cos\theta + 2\sin\theta \equiv \sqrt{20}\cos(\theta - 26.6°)$
(b) $\theta = 103.6°, 309.5°$
(c) Maximum $= \sqrt{20}$, minimum $= -\sqrt{20}$

2. (a) $\sqrt{89}\sin(x - 1.01219...)$
(b) $x = 1.70, 3.46$ (2 d.p.)

17 Trig modelling

G **1.** (a) $6\cos\theta + 8\sin\theta \equiv R(\cos\theta\cos\alpha + \sin\theta\sin\alpha)$
$R\cos\alpha = 6, R\sin\alpha = 8, \tan\alpha = \tfrac{8}{6}, \alpha = 0.92729...$
$R = \sqrt{6^2 + 8^2} = \sqrt{100} = 10$
So $6\cos\theta + 8\sin\theta \equiv 10\cos(\theta - 0.92729...)$
(b) (i) Maximum value $= 10$ (ii) $\theta = 0.92729...$
(c) Maximum temp. of 24°C at $t = 3.7$ hours
Minimum temp. of 4°C at $t = 16.3$ hours
(d) Temp. of 12°C when $t = 10.797...$ i.e. at 10.48 hours
i.e. 10.48 pm
and when $t = 21.753...$ i.e. at 21.45 hours i.e. 9.45 am
(the next day)
(e)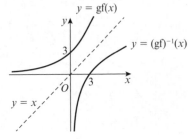

18 Exponential functions

G **1.** (a) $gf(x) = e^{(2x + \ln 3)} = e^{2x} \times e^{\ln 3} = 3e^{2x}$
Range of $gf(x)$ is $gf(x) > 0$

(b) $(gf)^{-1}(x) = \tfrac{1}{2}\ln\dfrac{x}{3}$ or $\ln\sqrt{\dfrac{x}{3}}$

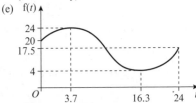

(c) $f^{-1}(x) = \tfrac{1}{2}(x - \ln 3)$
Domain is $x \in \mathbb{R}$

2. (a) $f^{-1}(x) = 1.5 - 0.25 e^x$
Domain is $x \in \mathbb{R}$ Range is $f^{-1}(x) < 1.5$
(b)

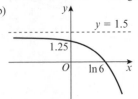

19 Exponential equations

G **1.** $\ln(5x + 24) = \ln(x + 2)^2$
$5x + 24 = (x + 2)^2$
$5x + 24 = x^2 + 4x + 4$
$0 = x^2 - x - 20$
$0 = (x - 5)(x + 4)$
$x = 5$
($x = -4$ is inadmissible since $\ln x$ is only defined for $x > 0$)

2. $x = 9$
($x = -2$ is inadmissible since $\ln x$ is only defined for $x > 0$)

3. $x = 0, x = \ln 2$

4. $x = \dfrac{\ln 5 + 1}{\ln 3 + 2}$

5. (a) $\dfrac{5x^2 - 13x - 6}{x^2 - 9} = \dfrac{(5x + 2)(x - 3)}{(x - 3)(x + 3)} = \dfrac{5x + 2}{x + 3}$

(b) $x = \dfrac{3e^2 - 2}{5 - e^2}$

20 Exponential modelling

1. (a) 370 °C

G (b) $280 = 350 e^{-0.08t} + 20$

$$\dfrac{260}{350} = e^{-0.08t}$$

$$\ln\left(\dfrac{260}{350}\right) = -0.08t$$

$$t = 3.72 \text{ minutes (3 s.f.)}$$

(c) 1.14 °C/min
(d) $e^{-0.08t} > 0$ for all values of t and $e^{-0.08t} \to 0$ as $t \to \infty$
So $T > 20$, i.e. the temperature can never fall to 18 °C

2. (a) 60 g
(b) $k = 0.00788$ (3 s.f.)
(c) 23.3 g (3 s.f.)
(d)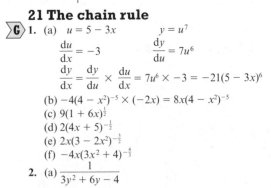

21 The chain rule

G **1.** (a) $u = 5 - 3x$ $y = u^7$

$$\dfrac{du}{dx} = -3 \qquad\qquad \dfrac{dy}{du} = 7u^6$$

$$\dfrac{dy}{dx} = \dfrac{dy}{du} \times \dfrac{du}{dx} = 7u^6 \times -3 = -21(5 - 3x)^6$$

(b) $-4(4 - x^2)^{-5} \times (-2x) = 8x(4 - x^2)^{-5}$
(c) $9(1 + 6x)^{\frac{1}{2}}$
(d) $2(4x + 5)^{-\frac{1}{2}}$
(e) $2x(3 - 2x^2)^{-\frac{1}{2}}$
(f) $-4x(3x^2 + 4)^{-\frac{4}{3}}$

2. (a) $\dfrac{1}{3y^2 + 6y - 4}$

(b) $-\dfrac{1}{4}$

3. $f'(x) = \dfrac{6(4\sqrt{x} + 3)^2}{\sqrt{x}}$

22 Derivatives to learn

G **1.** (a) $u = \sin x$ $y = u^3$

$$\dfrac{du}{dx} = \cos x \qquad\qquad \dfrac{dy}{du} = 3u^2$$

$$\dfrac{dy}{dx} = \dfrac{dy}{du} \times \dfrac{du}{dx} = 3u^2 \times \cos x = 3\sin^2 x \cos x$$

(b) $4\sin(5-4x)$

(c) $2x\,e^{x^2+1}$

(d) $\dfrac{3x^2}{x^3+2}$

2. (a) $-4\cos^3 x \sin x$

(b) $2\cos 2x - 3\sin 3x$

(c) $6x^2 + 3e^{5-3x}$

(d) $-2\tan 2x$

3. (a) $y = \operatorname{cosec} x = (\sin x)^{-1}$

$$\frac{dy}{dx} = -1(\sin x)^{-2}(\cos x) = \frac{-\cos x}{\sin^2 x} = \frac{-\cos x}{\sin x} \times \frac{1}{\sin x}$$
$$= -\cot x \operatorname{cosec} x$$

(b) $-\cot x$

23 The product rule

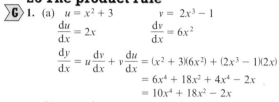

1. (a) $u = x^2 + 3$ $v = 2x^3 - 1$

$$\frac{du}{dx} = 2x \qquad\qquad \frac{dv}{dx} = 6x^2$$

$$\frac{dy}{dx} = u\frac{dv}{dx} + v\frac{du}{dx} = (x^2+3)(6x^2) + (2x^3-1)(2x)$$
$$= 6x^4 + 18x^2 + 4x^4 - 2x$$
$$= 10x^4 + 18x^2 - 2x$$

(b) $u = x^4$ $v = \sin x$

$$\frac{du}{dx} = 4x^3 \qquad\qquad \frac{dv}{dx} = \cos x$$

$$\frac{dy}{dx} = x^4(\cos x) + \sin x(4x^3)$$

(c) $4x\,e^{2x}(x+1)$

(d) $x^2 - 4 + (3x^2 - 4)\ln 2x$

2. (a) $h'(x) = e^{4x}\sec x\,(4 + \tan x)$

(b) $x = -1.33$ (3 s.f.)

3. $y = x^2(1+x^2)^{\frac{1}{2}}$

$$\frac{dy}{dx} = x^2\left[\tfrac{1}{2}(1+x^2)^{-\frac{1}{2}}(2x)\right] + (1+x^2)^{\frac{1}{2}}(2x)$$

$$= \frac{x^3}{(1+x^2)^{\frac{1}{2}}} + 2x(1+x^2)^{\frac{1}{2}}$$

$$= \frac{x^3 + 2x(1+x^2)}{(1+x^2)^{\frac{1}{2}}}$$

$$= \frac{3x^3 + 2x}{\sqrt{1+x^2}}$$

4. $(-1, 2e^{-2})$ and $(2, -e^4)$

24 The quotient rule

1. (a) $u = x^2$ $v = 2x+1$

$$\frac{du}{dx} = 2x \qquad\qquad \frac{dv}{dx} = 2$$

$$\frac{dy}{dx} = \frac{v\dfrac{du}{dx} - u\dfrac{dv}{dx}}{v^2} = \frac{(2x+1)(2x)-(x^2)(2)}{(2x+1)^2} = \frac{2x^2+2x}{(2x+1)^2}$$

(b) $u = x^3$ $v = (1-2x^2)^{\frac{1}{2}}$

$$\frac{du}{dx} = 3x^2 \qquad \frac{dv}{dx} = \tfrac{1}{2}(1-2x^2)^{-\frac{1}{2}}(-4x)$$
$$= (-2x)(1-2x^2)^{-\frac{1}{2}}$$

$$\frac{dy}{dx} = \frac{(1-2x^2)^{\frac{1}{2}}(3x^2)-(x^3)(-2x)(1-2x^2)^{-\frac{1}{2}}}{1-2x^2}$$

$$= \frac{(1-2x^2)(3x^2)-(x^3)(-2x)}{(1-2x^2)^{\frac{3}{2}}} = \frac{3x^2-4x^4}{(1-2x^2)^{\frac{3}{2}}}$$

2. (a) $\dfrac{x^2+1}{(1-x^2)^2}$

(b) $\dfrac{4}{(3x^2+1)^{\frac{3}{2}}}$

3. (a) $\dfrac{1-\ln 2x}{3x^2}$

(b) $\dfrac{-3}{1+\sin 3x}$

(c) $\dfrac{8\cos x - \sin x}{e^{2x}}$

25 Differentiation and graphs

1. (a) $\dfrac{dy}{dx} = (3x^2)\left(\dfrac{1}{x}\right) + (\ln x)(6x) = 3x + 6x\ln x$

(b) When $x = e$, $y = 3e^2\ln e = 3e^2$

When $x = e$, $\dfrac{dy}{dx} = 3e + 6e\ln e = 3e + 6e = 9e$

Equation of tangent is $y - 3e^2 = 9e(x-e)$,

i.e. $y = 9ex - 6e^2$

(c) $\left(\dfrac{2e}{3}, 0\right)$

2. (a) $y = e^{2x}\cos 3x$

$$\frac{dy}{dx} = e^{2x}(-3\sin 3x) + \cos 3x\,(2e^{2x})$$
$$= e^{2x}(2\cos 3x - 3\sin 3x)$$

Turnings points occur when $\dfrac{dy}{dx} = 0$,

i.e. $2\cos 3x = 3\sin 3x$, i.e. $\tan 3x = \tfrac{2}{3}$ (since $e^{2x} \neq 0$)

(b) $x + 2y = 2$

3. (a) Maximum at $(2, 1)$, minimum at $(-2, -1)$

(b) $12x - 25y + 8 = 0$

26 Iteration

1. (a) $f(2) = e^{-2} - 3 + 2\sqrt{2} = -0.0362...$

$f(3) = e^{-3} - 3 + 2\sqrt{3} = 0.5138...$

Change of sign, so root between 2 and 3

(b) $e^{-x} - 3 + 2\sqrt{x} = 0$

$$2\sqrt{x} = 3 - e^{-x}$$
$$\sqrt{x} = \tfrac{1}{2}(3 - e^{-x})$$
$$x = \tfrac{1}{4}(3 - e^{-x})^2$$

(c) $x_1 = \tfrac{1}{4}(3-e^{-2.1})^2 = 2.0701$, $x_2 = 2.0647$, $x_3 = 2.0637$

(all to 4 d.p.)

(d) $f(2.0635) = -0.000013790$, $f(2.0645) = 0.000555323$

Change of sign, so $2.0635 < \alpha < 2.0645$

i.e. $\alpha = 2.064$ (3 d.p.)

2. (a) $f(1) = -2$ and $f(2) = 10$

Change of sign, so root between 1 and 2

(b) $x^3 + 3x^2 - 4x - 2 = 0$

$$x^2(x+3) = 4x + 2$$
$$x^2 = \frac{4x+2}{x+3}$$
$$x = \sqrt{\frac{4x+2}{x+3}}$$

(c) $x_1 = 1.3333$, $x_2 = 1.3009$, $x_3 = 1.2942$ (all to 4 d.p.)

(d) $f(1.29235) = -0.000452142$, $f(1.29245) = 0.000424387$

Change of sign, so $1.29235 < \alpha < 1.29245$

i.e. $\alpha = 1.2924$ (4 d.p.)

27–28 You are the examiner!

1. $\dfrac{5x}{(x+4)(x-6)} - \dfrac{3}{x-6} + \dfrac{2}{x+4}$

$$= \frac{5x - 3x - 12 + 2x - 12}{(x+4)(x-6)} = \frac{4x-24}{(x+4)(x-6)}$$

$$= \frac{4(x-6)}{(x+4)(x-6)} = \frac{4}{x+4}$$

2. (a) $3\sin\theta + 7\cos\theta = R\sin\theta\cos\alpha + R\cos\theta\sin\alpha$

$R\cos\alpha = 3$, $R\sin\alpha = 7$, $\tan\alpha = \tfrac{7}{3}$, $\alpha = 66.8°$

$R = \sqrt{3^2 + 7^2} = \sqrt{58}$

So $3\sin\theta + 7\cos\theta = \sqrt{58}\sin(\theta + 66.8°)$

(b) $\sqrt{58}\sin(\theta + 66.8°) = 4$, $\sin(\theta + 66.8°) = \dfrac{4}{\sqrt{58}}$

where $66.8° \leq (\theta + 66.8°) \leq 426.8°$

$\theta + 66.8° = 148.3°$ or $391.7°$

$\theta = 81.5°$ or $324.9°$

3. (a) $f(x) = \ln(2x+3)$

Let $y = \ln(2x+3)$, $e^y = 2x+3$, $x = \dfrac{e^y - 3}{2}$

So $f^{-1}(x) = \dfrac{e^x - 3}{2}$

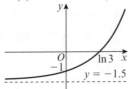

(b) When $x = 0$, $f^{-1}(x) = -1$ and since $e^x > 0$ for all x,
$f^{-1}(x) > -1.5$ (asymptote at $y = -1.5$)
$f^{-1}(x) = 0$ when $e^x = 3$, i.e. $x = \ln 3$

4. $y = \dfrac{4 - 3x}{e^{2x}}$

$\dfrac{dy}{dx} = \dfrac{e^{2x}(-3) - (4 - 3x)(2e^{2x})}{(e^{2x})^2} = \dfrac{e^{2x}(-3 - 8 + 6x)}{(e^{2x})^2}$

$= \dfrac{6x - 11}{e^{2x}}$

5. (a) $\ln(3x - 2) + \ln(x - 2) = 2\ln(x + 2)$
$\ln[(3x - 2)(x - 2)] = \ln[(x + 2)^2]$
$3x^2 - 8x + 4 = x^2 + 4x + 4$
$2x^2 - 12x = 0$
$2x(x - 6) = 0$
$x = 6$ ($x = 0$ inadmissible, since $x > 2$)
(b) $2e^x + 5 = 3e^{-x}$
$2(e^x)^2 + 5e^x - 3 = 0$
$(2e^x - 1)(e^x + 3) = 0$
$e^x = 0.5$ or $e^x = -3$ (inadmissible, since $e^x > 0$)
$x = \ln(0.5)$ or $-\ln 2$

Core Mathematics 4
29 Partial fractions

G 1. $\dfrac{3x - 10}{(x - 2)(x - 4)} = \dfrac{A}{x - 2} + \dfrac{B}{x - 4}$

When $x = 2$ work out $\dfrac{3x - 10}{x - 4}$: $A = \dfrac{-4}{-2} = 2$

When $x = 4$ work out $\dfrac{3x - 10}{x - 2}$: $B = \dfrac{2}{2} = 1$

So $\dfrac{3x - 10}{(x - 2)(x - 4)} \equiv \dfrac{2}{x - 2} + \dfrac{1}{x - 4}$

2. $\dfrac{2}{2x + 1} + \dfrac{1}{4x - 1}$

3. $-\dfrac{1}{x} + \dfrac{4}{5(x + 2)} + \dfrac{1}{5(x - 3)}$

G 4. $x^2 - 13 = A(x - 1)(x + 2) + B(x + 2) + C(x - 1)^2$
When $x = -2$, $-9 = 9C$, so $C = -1$
When $x = 1$, $-12 = 3B$, so $B = -4$
Comparing coefficients of x^2, $1 = A + C$, so $A = 2$

5. $A = 1$, $B = -2$, $C = 1$, $D = -1$

30 Parametric equations

G 1. $t = 1 - x$, so $t^2 = (1 - x)^2 = 1 - 2x + x^2$
$y = t^2 - 4 = 1 - 2x + x^2 - 4$ so $y = x^2 - 2x - 3$

2. (a) $xy^2 = 9$
(b) $y = x^2 + 2x + 2$
(c) $25x^2 - y^2 = 25$
(d) $4x^2 + y^2 - 8x - 2y + 1 = 0$

3. $(-\frac{4}{3}, -3)$ and $(1, \frac{1}{2})$

4. $2x^2 - 2xy + 5y^2 = 9$

31 Parametric differentiation

G 1. (a) $\dfrac{dx}{dt} = -4\sin t$, $\dfrac{dy}{dt} = 3\cos t$, $\dfrac{dy}{dx} = \dfrac{dy}{dt} \div \dfrac{dx}{dt}$

$= \dfrac{3\cos t}{-4\sin t} = -\dfrac{3}{4\tan t}$ or $-\dfrac{3}{4}\cot t$

(b) $\sqrt{3}x + 4y = 8\sqrt{3}$ or $3x + 4\sqrt{3}y = 24$

2. (a) $-\frac{1}{2}e^{4t}$
(b) -8
(c) $(-2, 3)$
(d) $(-26, 0)$
(e) $xy + 3y + 5x = -7$

32 The binomial series

G 1. $(1 + 4x)^{\frac{3}{2}} = 1 + (\frac{3}{2})(4x) + \dfrac{(\frac{3}{2})(\frac{1}{2})(4x)^2}{1 \times 2} + \dfrac{(\frac{3}{2})(\frac{1}{2})(-\frac{1}{2})(4x)^3}{1 \times 2 \times 3} \cdots$

$= 1 + 6x + 6x^2 - 4x^3 \cdots$

G 2. $\sqrt{4 - x} = (4 - x)^{\frac{1}{2}} = \left[4\left(1 - \dfrac{x}{4}\right)\right]^{\frac{1}{2}} = 4^{\frac{1}{2}}\left(1 - \dfrac{x}{4}\right)^{\frac{1}{2}} = 2\left(1 - \dfrac{x}{4}\right)^{\frac{1}{2}}$

$2\left(1 - \dfrac{x}{4}\right)^{\frac{1}{2}} = 2\left[1 + \left(\frac{1}{2}\right)\left(\dfrac{-x}{4}\right) + \dfrac{\left(\frac{1}{2}\right)\left(-\frac{1}{2}\right)\left(\dfrac{-x}{4}\right)^2}{1 \times 2}\right.$

$\left. + \dfrac{\left(\frac{1}{2}\right)\left(\frac{-1}{2}\right)\left(\frac{-3}{2}\right)\left(\dfrac{-x}{4}\right)^3}{1 \times 2 \times 3} \cdots\right]$

$= 2\left(1 - \dfrac{x}{8} - \dfrac{x^2}{128} - \dfrac{x^3}{1024} \cdots\right)$

$= 1 - \dfrac{x}{4} - \dfrac{x^2}{64} - \dfrac{x^3}{512} \cdots$

3. (a) $\frac{1}{2} - \dfrac{5x}{16} + \dfrac{75x^2}{256} - \dfrac{625x^3}{2048} \cdots$
(b) 0.50637 (5 d.p.) (using $x = -0.02$)
(c) $2 - \dfrac{3x}{4} + \dfrac{55x^2}{64} \cdots$

33 Implicit differentiation

G 1. Differentiating implicitly, $2x + 4\dfrac{dy}{dx} - 2y\dfrac{dy}{dx} = 0$

$2x = (2y - 4)\dfrac{dy}{dx}$, $\dfrac{dy}{dx} = \dfrac{2x}{2y - 4} = \dfrac{x}{y - 2}$

When $x = -2$ and $y = 3$, $\dfrac{dy}{dx} = -2$,

So gradient of normal $= \frac{1}{2}$
Equation of normal is $y - 3 = \frac{1}{2}(x - -2)$ i.e. $x - 2y + 8 = 0$

2. $(2, -4)$ and $(-2, 4)$

3. (a) $\dfrac{2ye^{2x}}{2\sin 2y - e^{2x}}$

(b) $-\pi$
(c) $\ln 2 + \frac{1}{4}$

34 Differentiating a^x

G 1. Let $u = 2x$, $\dfrac{du}{dx} = 2$, $y = 4^u$, $\dfrac{dy}{du} = 4^u \ln 4$

$\dfrac{dy}{dx} = \dfrac{dy}{du} \times \dfrac{du}{dx} = 4^u \ln 4 \times 2 = 2 \times 4^{2x} \times \ln 4$

(simplifies to $2^{4x+1} \ln 4$ or $4^{2x+1} \ln 2$)

2. (a) $3^{\tan x} \ln 3 \sec^2 x$
(b) $6^{2x}x^2(3 + 2x\ln 6)$
(c) $\dfrac{5^{x^2}}{x} + 2x\, 5^{x^2} \ln 5 \ln 2x$

3. $y = \left(\frac{9}{4}\ln 3\right)x + 2$

4. (a) $(2, -4)$ and $\left(2, \frac{4}{3}\right)$

(b) At $(2, -4)$, $\dfrac{dy}{dx} = -\ln 4 - 1$

At $\left(2, \frac{4}{3}\right)$, $\dfrac{dy}{dx} = \ln 4 - \frac{1}{3}$

35 Rates of change

G 1. $\dfrac{dV}{dt} = 8$, $V = \frac{2}{9}\pi x^3$, so $\dfrac{dV}{dx} = \frac{2}{3}\pi x^2$

Rate required $= \dfrac{dx}{dt} = \dfrac{dx}{dV} \times \dfrac{dV}{dt} = \dfrac{dV}{dt} \div \dfrac{dV}{dx}$

$= 8 \div \frac{2}{3}\pi x^2 = \dfrac{12}{\pi x^2}$

When $x = 6$, $\dfrac{dx}{dt} = \dfrac{1}{3\pi}$ cm s^{-1}

2. 0.0249 cm s^{-1}

3. (a) Volume of water = cross-sectional area $\times$ length
$= \frac{1}{2}xh \times 4 = 2xh$
Using similar triangles, $\dfrac{1.5}{1} = \dfrac{x}{h}$, i.e. $x = 1.5h$,
so volume of water, $V = 2 \times 1.5h \times h = 3h^2$
(b) 0.01 m s^{-1}

36 Integrals to learn

G 1. (a) $\int 6\cos 2x \, dx = 3\sin 2x + c$
(b) $2\tan \dfrac{x}{2} + c$
(c) $-\frac{1}{6}\cos 3x + c$
(d) $-\frac{1}{2}\cosec 2\theta + c$

2. (a) $3\cos 4x + c$ (b) $8\sin\dfrac{\theta}{2} + c$
(c) $-\frac{1}{2}\cos 10x + c$ (d) $\frac{1}{6}\sin(3x + 5) + c$

3. (a) $\frac{2}{3}e^{3x} + c$

 (b) $\frac{12}{5}e^{\frac{x}{4}} + c$

 (c) $\frac{5}{2}\ln(2x - 1) + c$

 (d) $\frac{1}{4}\ln(1 + 8x) + c$

4. (a) $\frac{2}{3}(x + 6)^{\frac{3}{2}} + c$

 (b) $\frac{1}{6}(2x - 3)^{\frac{3}{2}} + c$

 (c) $-\frac{2}{15}(4 - 3x)^{\frac{5}{2}} + c$

 (d) $\frac{3}{16}(1 + 4x)^{\frac{4}{3}} + c$

5. $\int_2^6 \frac{1}{\sqrt{5x + 2}}\,dx = \int_2^6 (5x + 2)^{-\frac{1}{2}}dx = \frac{2}{5}\Big[(5x + 2)^{\frac{1}{2}}\Big]_2^6$

 $= \frac{2}{5}(\sqrt{32} - \sqrt{12}) = \frac{2}{5}(4\sqrt{2} - 2\sqrt{3})$

 $= \frac{4}{5}(2\sqrt{2} - \sqrt{3})$

6. $\frac{9}{14}$

37 Reverse chain rule

1. (a) $\int \frac{3x - 4}{3x^2 - 8x}\,dx = \frac{1}{2}\ln(3x^2 - 8x) + c$

 (b) $\int 6x^2(7 - 4x^3)^6\,dx = -\frac{1}{14}(7 - 4x^3)^7 + c$

2. (a) $\frac{1}{72}(6x^2 - 2)^6 + c$

 (b) $-\frac{3}{2}\ln(\cos 2\theta) + c$

3. (a) $-\frac{3}{4}\cos^4 x + c$

 (b) $\frac{1}{6}\sin(x^2) + c$

4. (a) $\frac{1}{2}\tan^4\theta + c$

 (b) $-\frac{3}{5}\operatorname{cosec}^5 x + c$

5. (a) $\frac{4}{9}(3x^2 - 2)^{\frac{3}{2}} + c$

 (b) $\frac{2}{3}(2x^3 - 6x^2 + 5)^{\frac{1}{2}} + c$

6. $\ln 3$

7. $\ln(1.5)$

8. 0.576 (3 d.p.)

38 Integrating partial fractions

1. (a) $\dfrac{5 - 8x}{(2 + x)(1 - 3x)} = \dfrac{A}{2 + x} + \dfrac{B}{1 - 3x}$

 so $5 - 8x = A(1 - 3x) + B(2 + x)$

 When $x = -2$, $21 = 7A$, so $A = 3$

 When $x = \frac{1}{3}$, $\frac{7}{3} = \frac{7}{3}B$ so $B = 1$

 So $\dfrac{5 - 8x}{(2 + x)(1 - 3x)} \equiv \dfrac{3}{2 + x} + \dfrac{1}{1 - 3x}$

 (b) $\int_{-1}^0 \dfrac{5 - 8x}{(2 + x)(1 - 3x)}\,dx = \int_{-1}^0 \left(\dfrac{3}{(2 + x)} + \dfrac{1}{(1 - 3x)}\right)dx$

 $= \Big[3\ln(2 + x) - \frac{1}{3}\ln(1 - 3x)\Big]_{-1}^0$

 $= (3\ln 2 - 0) - (0 - \frac{1}{3}\ln 4)$

 $= 3\ln 2 + \frac{2}{3}\ln 2 = \frac{11}{3}\ln 2$

2. (a) $-\dfrac{1}{2x + 3} + \dfrac{2}{2x - 3}$

 (b) $\dfrac{12x^3 - 31x - 18}{4x^2 - 9} = \dfrac{3x(4x^2 - 9) - 4x - 18}{4x^2 - 9}$

 $= 3x - \dfrac{4x + 18}{4x^2 - 9}$

 $= 3x - \dfrac{2(2x + 9)}{4x^2 - 9}$

 (c) $\frac{3}{2} + \ln 15$

39 Identities in integration

1. $\cos 2A = 2\cos^2 A - 1$

 When $A = \frac{x}{2}$, $\cos x = 2\cos^2\left(\frac{x}{2}\right) - 1$

 So $\cos^2\left(\frac{x}{2}\right) = \frac{1}{2}(\cos x + 1)$

 $\int_{\frac{\pi}{6}}^{\frac{\pi}{2}} \cos^2\left(\frac{x}{2}\right)dx = \int_{\frac{\pi}{6}}^{\frac{\pi}{2}} \frac{1}{2}(\cos x + 1)\,dx = \frac{1}{2}\Big[\sin x + x\Big]_{\frac{\pi}{6}}^{\frac{\pi}{2}}$

 $= \frac{1}{2}\Big[\left(1 + \frac{\pi}{2}\right) - \left(\frac{1}{2} + \frac{\pi}{6}\right)\Big] = \frac{1}{4} + \frac{\pi}{6}$

2. $\frac{1}{8}$

3. (a) $\cos 5x = \cos(3x + 2x) = \cos 3x\cos 2x - \sin 3x\sin 2x$

 $\cos x = \cos(3x - 2x) = \cos 3x\cos 2x + \sin 3x\sin 2x$

 So $\cos 5x + \cos x = 2\cos 3x\cos 2x$

 (b) $\frac{3}{10}$

4. $\dfrac{\pi}{8} - \dfrac{1}{4}$

5. $\dfrac{1}{\sqrt{3}} - \dfrac{\pi}{6}$ $\left(\text{can also be written as } \dfrac{\sqrt{3}}{3} - \dfrac{\pi}{6} \text{ or } \dfrac{2\sqrt{3} - \pi}{6}\right)$

40 Integration by substitution

1. $u = \sin 2x$, $\dfrac{du}{dx} = 2\cos 2x$ so $\dfrac{du}{2} = \cos 2x\,dx$

 When $x = \dfrac{\pi}{12}$, $u = \sin\dfrac{\pi}{6} = 0.5$; when $x = \dfrac{\pi}{4}$, $u = \sin\dfrac{\pi}{2} = 1$

 $\int_{\frac{\pi}{12}}^{\frac{\pi}{4}} \sin^3 2x\cos 2x\,dx = \int_{0.5}^1 \frac{1}{2}u^3\,du = \Big[\frac{1}{8}u^4\Big]_{0.5}^1 = \frac{15}{128}$

2. $\dfrac{2\sqrt{3}}{27}$

3. $\dfrac{e - 1}{2(1 + e)}$

4. π

41 Integration by parts

1. $u = x$, $\dfrac{du}{dx} = 1$, $\dfrac{dv}{dx} = e^{-x}$, $v = -e^{-x}$

 $\int xe^{-x}\,dx = x(-e^{-x}) - \int(-e^{-x})(1)\,dx$

 $= -xe^{-x} + \int(e^{-x})\,dx = -xe^{-x} - e^{-x} + c$

 $= -e^{-x}(x + 1) + c$

2. $u = \ln x$, $\dfrac{du}{dx} = \dfrac{1}{x}$, $\dfrac{dv}{dx} = x^3$, $v = \dfrac{x^4}{4}$

 $\int x^3\ln x\,dx = \dfrac{x^4}{4}\ln x - \int\dfrac{x^4}{4}\dfrac{1}{x}\,dx$

 $= \dfrac{x^4}{4}\ln x - \int\dfrac{x^3}{4}\,dx$

 $= \dfrac{x^4}{4}\ln x - \dfrac{x^4}{16} + c$

 $= \dfrac{x^4}{16}(4\ln x - 1) + c$

3. (a) $-2x\cos x + 2\sin x + c$

 (b) $(x^2 - 2)\sin x + 2x\cos x + c$

4. $\frac{1}{4}e^2(5e^2 - 1)$

42 Areas and parametric curves

1. $x = 2t - \sin t$, $\dfrac{dx}{dt} = 2 - \cos t$, $y = 1 - \cos t$

 When $y = 0$, $\cos t = 1$, so $t = 0$ or 2π

 Area $= \int_{x=0}^{x=4\pi} y\,dx = \int_{t=0}^{t=2\pi} y\dfrac{dx}{dt}\,dt = \int_0^{2\pi}(1 - \cos t)(2 - \cos t)\,dt$

 $= \int_0^{2\pi}(2 - 3\cos t + \cos^2 t)\,dt$

 $= \int_0^{2\pi}\left(2 - 3\cos t + \frac{1}{2} + \frac{1}{2}\cos 2t\right)dt$

 $= \int_0^{2\pi}\left(-3\cos t + \frac{5}{2} + \frac{1}{2}\cos 2t\right)dt$

 $= \Big[-3\sin t + \frac{5t}{2} + \frac{1}{4}\sin 2t\Big]_0^{2\pi}$

 $= (0 + 5\pi + 0) - (0) = 5\pi$

2. $\frac{16}{3}$

3. $3e - 6$

43 Volumes of revolution 1

1. $V = \pi\int_0^2 y^2\,dx = \pi\int_0^2(x^3 - 2x^2)^2\,dx = \pi\int_0^2(x^6 - 4x^5 + 4x^4)\,dx$

 $= \pi\Big[\dfrac{x^7}{7} - \dfrac{2x^6}{3} + \dfrac{4x^5}{5}\Big]_0^2 = \pi\Big[\dfrac{128}{7} - \dfrac{128}{3} + \dfrac{128}{5} - (0)\Big] = \dfrac{128\pi}{105}$

2. $\frac{3}{8}\pi\ln 33$

3. $\dfrac{9\pi^2}{4}$

4. $11\pi^2$

44 Volumes of revolution 2

1. When $x = 0$, $t = 1$ and when $x = 3$, $t = 2$.

 Volume $= \pi\int_0^3 y^2\,dx = \pi\int_1^2 y^2\dfrac{dx}{dt}\,dt = \pi\int_1^2 t^6 2t\,dt = 2\pi\int_1^2 t^7\,dt$

 $= 2\pi\Big[\dfrac{t^8}{8}\Big]_1^2 = 2\pi\left(\dfrac{256}{8} - \dfrac{1}{8}\right) = \dfrac{255\pi}{4}$

2. (a) $(1, 3)$

 (b) $\dfrac{389}{35}\pi$

3. $\frac{1}{8}\pi(\pi^2 - 4)$

45 The trapezium rule

1. (a)

x	0	0.8	1.6	2.4	3.2	4
y	0	0.303 03	0.350 88	0.309 28	0.261 44	0.222 22

(b) $\int_0^4 \frac{x}{x^2+2}\,dx \approx \frac{0.8}{2}[0 + 0.222\,22 + 2(0.303\,03 + 0.350\,88$
$+ 0.309\,28 + 0.261\,44)] = 1.0686$ (4 d.p.)

(c) $\ln 3$

(d) 2.73% (3 s.f.)

2. (a)

x	1	1.5	2	2.5	3
y	0	0.810 93	2.079 44	3.665 16	5.493 06

(b) 4.6510 (4 d.p.)

(c) Let $A = \int_1^3 (2x-1)\ln x\,dx = \int_1^3 2x\ln x\,dx - \int_1^3 1\ln x\,dx$

$\int_1^3 2x\ln x\,dx = [x^2\ln x]_1^3 - \int_1^3 x^2\frac{1}{x}\,dx = \left[x^2\ln x - \frac{x^2}{2}\right]_1^3$

$\int_1^3 1\ln x\,dx = [x\ln x]_1^3 - \int_1^3 x\frac{1}{x}\,dx = [x\ln x - x]_1^3$

So $A = \left[x^2\ln x - \frac{x^2}{2} - x\ln x + x\right]_1^3$

$= (9\ln 3 - 4.5 - 3\ln 3 + 3) - (0 - 0.5 - 0 + 1)$

$= 6\ln 3 - 2$

46 Solving differential equations

1. (a) $\frac{10}{(x-3)(3x+1)} = \frac{A}{x-3} + \frac{B}{3x+1}$

i.e. $10 = A(3x+1) + B(x-3)$

When $x = 3$, $10 = 10A$, so $A = 1$

When $x = -\frac{1}{3}$, $10 = -\frac{10}{3}B$, so $B = -3$

Hence $\frac{10}{(x-3)(3x+1)} = \frac{1}{x-3} - \frac{3}{3x+1}$

(b) $\int \frac{10}{(x-3)(3x+1)}\,dx = \int\left(\frac{1}{x-3} - \frac{3}{3x+1}\right)dx$

$= \ln(x-3) - \ln(3x+1) + c = \ln\left(\frac{x-3}{3x+1}\right) + c$

(c) $y = \frac{32(x-3)}{3x+1}$ (obtained using $\ln k$ as a constant of integration)

2. $h = 4.10\,\text{m}$

47 Vectors

1. (a) $|\overrightarrow{OP}| = \sqrt{1^2 + 7^2 + 5^2} = \sqrt{1 + 49 + 25} = \sqrt{75} = 5\sqrt{3}$

(b) $\overrightarrow{QP} = \overrightarrow{QO} + \overrightarrow{OP} = \overrightarrow{OP} - \overrightarrow{OQ}$
$= \mathbf{i} + 7\mathbf{j} + 5\mathbf{k} - (3\mathbf{i} + 2\mathbf{j} - 6\mathbf{k}) = -2\mathbf{i} + 5\mathbf{j} + 11\mathbf{k}$

(c) $\sqrt{150}$ or $5\sqrt{6}$

2. (a) $\frac{1}{9}(8\mathbf{i} - 4\mathbf{j} - \mathbf{k})$

(b) $\frac{1}{7}(-6\mathbf{i} + 2\mathbf{j} + 3\mathbf{k})$

(c) $\frac{1}{9}(8\mathbf{i} - 4\mathbf{j} - \mathbf{k})$

3. (a) (i) $\overrightarrow{PQ} = -3\mathbf{i} - 5\mathbf{j} + \mathbf{k}$ (ii) $|\overrightarrow{PQ}| = \sqrt{35}$

(b) (i) $\overrightarrow{PQ} = -6\mathbf{i} - 2\mathbf{j} + 9\mathbf{k}$ (ii) $|\overrightarrow{PQ}| = 11$

4. $\lambda = 2$ and $\lambda = -2$

5. $\mu = 2$ and $\mu = -\frac{12}{5}$

48 Vector equations of lines

1. (a) $\overrightarrow{PQ} = \overrightarrow{PO} + \overrightarrow{OQ} = \overrightarrow{OQ} - \overrightarrow{OP}$
$= 3\mathbf{i} - 3\mathbf{j} + 4\mathbf{k} - (\mathbf{i} + 2\mathbf{j} - \mathbf{k})$
$= 2\mathbf{i} - 5\mathbf{j} + 5\mathbf{k}$

(b) Line L has equation $\mathbf{r} = \mathbf{i} + 2\mathbf{j} - \mathbf{k} + \lambda(2\mathbf{i} - 5\mathbf{j} + 5\mathbf{k})$

2. (a) $\mathbf{r} = 2\mathbf{i} - 3\mathbf{j} + \mathbf{k} + 4\lambda\mathbf{j}$ or $\mathbf{r} = 2\mathbf{i} + \mathbf{j} + \mathbf{k} + 4\lambda\mathbf{j}$

(b) $\mathbf{r} = \mathbf{i} + 5\mathbf{j} - 6\mathbf{k} + \lambda(3\mathbf{i} - 7\mathbf{j} + 5\mathbf{k})$ or
$\mathbf{r} = 4\mathbf{i} - 2\mathbf{j} - \mathbf{k} + \lambda(3\mathbf{i} - 7\mathbf{j} + 5\mathbf{k})$

3. $a = 26$ and $b = -7$

4. (a) $a = -29$ and $b = 7$

(b) $(1, 2, -19)$

49 Intersecting lines

1. $\begin{pmatrix} 4 \\ -5 \\ 3 \end{pmatrix} + \lambda\begin{pmatrix} -1 \\ 3 \\ 1 \end{pmatrix} = \begin{pmatrix} 7 \\ -8 \\ 6 \end{pmatrix} + \mu\begin{pmatrix} 2 \\ -3 \\ 1 \end{pmatrix}$

$4 - \lambda = 7 + 2\mu$ ①
$-5 + 3\lambda = -8 - 3\mu$ ②
$3 + \lambda = 6 + \mu$ ③

① + ③: $7 = 13 + 3\mu$, so $\mu = -2$, hence $\lambda = 1$
Substitute $\lambda = 1$ and $\mu = -2$ into ②:
LHS $= -5 + 3 = -2$, RHS $= -8 + 6 = -2$
Using L_1, position vector of point of intersection is

$\begin{pmatrix} 4 \\ -5 \\ 3 \end{pmatrix} + 1\begin{pmatrix} -1 \\ 3 \\ 1 \end{pmatrix} = \begin{pmatrix} 3 \\ -2 \\ 4 \end{pmatrix}$

2. (a) The direction vector of L_1 is $\begin{pmatrix} 1 \\ -2 \\ 1 \end{pmatrix}$,

the direction vector of L_2 is $\begin{pmatrix} -2 \\ 4 \\ -2 \end{pmatrix} = -2\begin{pmatrix} 1 \\ -2 \\ 1 \end{pmatrix}$

and since one is a multiple of the other, the lines must be parallel.

(b) $\overrightarrow{OA} = \begin{pmatrix} 1 \\ -3 \\ 5 \end{pmatrix}$ and $\overrightarrow{OB} = \begin{pmatrix} 11 \\ -15 \\ 6 \end{pmatrix}$

$\overrightarrow{AB} = \overrightarrow{AO} + \overrightarrow{OB} = \begin{pmatrix} -1 \\ 3 \\ -5 \end{pmatrix} + \begin{pmatrix} 11 \\ -15 \\ 6 \end{pmatrix} = \begin{pmatrix} 10 \\ -12 \\ 1 \end{pmatrix}$

Equation of line through A and B is $\mathbf{r} = \begin{pmatrix} 1 \\ -3 \\ 5 \end{pmatrix} + t\begin{pmatrix} 10 \\ -12 \\ 1 \end{pmatrix}$

When $t = -2$ this gives $\begin{pmatrix} -19 \\ 21 \\ 3 \end{pmatrix}$ which is the position

vector of C, so all three points are collinear.

50 Scalar product

1. $\overrightarrow{OP}.\overrightarrow{OQ} = (-3)(2) + (5)(4) + (-2)(1) = 12$

$|\overrightarrow{OP}| = \sqrt{(-3)^2 + 5^2 + (-2)^2} = \sqrt{38}$

$|\overrightarrow{OQ}| = \sqrt{2^2 + 4^2 + 1^2} = \sqrt{21}$

$\cos POQ = \frac{\overrightarrow{OP}.\overrightarrow{OQ}}{|\overrightarrow{OP}||\overrightarrow{OQ}|} = \frac{12}{\sqrt{38}\sqrt{21}}$

so angle $POQ = 64.9°$ (1 d.p.)

2. $59.7°$ (1 d.p.)

3. (a) $\overrightarrow{DE} = \begin{pmatrix} -6 \\ 3 \\ -2 \end{pmatrix}$

(b) $80.1°$ (1 d.p.)

51 Perpendicular vectors

1. $\overrightarrow{BA} = \overrightarrow{BO} + \overrightarrow{OA} = -2\mathbf{i} + \mathbf{j} - 3\mathbf{k} + 5\mathbf{i} + 3\mathbf{j} + 2\mathbf{k}$
$= 3\mathbf{i} + 4\mathbf{j} - \mathbf{k}$
$\overrightarrow{BC} = \overrightarrow{BO} + \overrightarrow{OC} = -2\mathbf{i} + \mathbf{j} - 3\mathbf{k} + 7\mathbf{i} - 3\mathbf{j} + 10\mathbf{k}$
$= 5\mathbf{i} - 2\mathbf{j} + 7\mathbf{k}$
$\overrightarrow{BA}.\overrightarrow{BC} = (3)(5) + (4)(-2) + (-1)(7) = 15 - 8 - 7 = 0$,
so $\angle ABC = 90°$

2. (a) $(5, -1, -2)$

(b) $42.7°$ (1 d.p.)

(c) $(11, -10, -20)$

(d) $\overrightarrow{AB} = \begin{pmatrix} 6 \\ -9 \\ -18 \end{pmatrix}$

(e) $|\overrightarrow{AB}| = 21$

(f) 28.6 (1 d.p.)

52 Solving area problems

1. $|\overrightarrow{BA}| = \sqrt{2^2 + 4^2 + (-5)^2} = \sqrt{45}$

$|\overrightarrow{BC}| = \sqrt{4^2 + 2^2 + (-6)^2} = \sqrt{56}$

$\cos ABC = \frac{\overrightarrow{BA}.\overrightarrow{BC}}{|\overrightarrow{BA}||\overrightarrow{BC}|} = \frac{(2)(4) + (4)(2) + (-5)(-6)}{\sqrt{45}\sqrt{56}} = \frac{46}{\sqrt{45}\sqrt{56}}$

so $\angle ABC = 23.6°$

Area of $\triangle ABC = \frac{1}{2} \times |\overrightarrow{BA}| \times |\overrightarrow{BC}| \times \sin ABC$
$= \frac{1}{2} \times \sqrt{45} \times \sqrt{56} \times \sin 23.6° = 10.05$ (2 d.p.)

2. (a) $\overrightarrow{AB} = \begin{pmatrix} -2 \\ 4 \\ -4 \end{pmatrix}$

(b) 60.9° (1 d.p.)

(c) $(-3, 6, 6)$

(d) 32.3 (1 d.p.)

(e) $(-1, 2, 10)$

53–54 You are the examiner!

1. $x - 5 \equiv A(x - 1) + B(x + 1)(x - 1) + C(x + 1)^2$

When $x = 1$, $-4 = 4C$, so $C = -1$

When $x = -1$, $-6 = -2A$, so $A = 3$

Comparing coefficients of x^2, $0 = B + C$, so $B = 1$

So $\dfrac{x - 5}{(x + 1)^2(x - 1)} \equiv \dfrac{3}{(x + 1)^2} + \dfrac{1}{x + 1} - \dfrac{1}{x - 1}$

2. Let $I = \int_{\frac{\pi}{4}}^{\frac{\pi}{2}} 5x \sin 2x \, dx$

$u = 5x, \dfrac{du}{dx} = 5, \dfrac{dv}{dx} = \sin 2x, v = -\tfrac{1}{2}\cos 2x$

$I = 5x(-\tfrac{1}{2}\cos 2x) - \int 5(-\tfrac{1}{2}\cos 2x)\,dx$

$= \left[5x(-\tfrac{1}{2}\cos 2x) + \tfrac{5}{4}\sin 2x\right]_{\frac{\pi}{4}}^{\frac{\pi}{2}}$

$= \left(\dfrac{5\pi}{4} + 0\right) - \left(0 + \dfrac{5}{4}\right) = \dfrac{5(\pi - 1)}{4}$

3. (a) $y^3 - xy^2 - x^3 = 3$

$3y^2 \dfrac{dy}{dx} - \left(x 2y \dfrac{dy}{dx} + y^2\right) - 3x^2 = 0$

$\dfrac{dy}{dx} = \dfrac{3x^2 + y^2}{3y^2 - 2xy}$

(b) At $(1, 2)$, $\dfrac{dy}{dx} = \dfrac{7}{8}$, so gradient of normal $= -\dfrac{8}{7}$

4. (a) $\overrightarrow{PQ} = \overrightarrow{PO} + \overrightarrow{OQ} = (-2\mathbf{i} + 5\mathbf{j} + \mathbf{k}) + (4\mathbf{i} + \mathbf{j} - 2\mathbf{k})$

$= 2\mathbf{i} + 6\mathbf{j} - \mathbf{k}$

Equation of L_1 is $\mathbf{r} = \begin{pmatrix} 2 \\ -5 \\ -1 \end{pmatrix} + \lambda \begin{pmatrix} 2 \\ 6 \\ -1 \end{pmatrix}$ or

$\mathbf{r} = \begin{pmatrix} 4 \\ 1 \\ -2 \end{pmatrix} + \lambda \begin{pmatrix} 2 \\ 6 \\ -1 \end{pmatrix}$

(b) We want the acute angle between $\begin{pmatrix} 2 \\ 6 \\ -1 \end{pmatrix}$ and $\begin{pmatrix} 0 \\ 1 \\ -3 \end{pmatrix}$

$\begin{pmatrix} 2 \\ 6 \\ -1 \end{pmatrix}$ has magnitude $\sqrt{41}$, $\begin{pmatrix} 0 \\ 1 \\ -3 \end{pmatrix}$ has magnitude $\sqrt{10}$,

$\begin{pmatrix} 2 \\ 6 \\ -1 \end{pmatrix} \cdot \begin{pmatrix} 0 \\ 1 \\ -3 \end{pmatrix} = 0 + 6 + 3 = 9$

$\cos \theta = \dfrac{9}{\sqrt{41}\sqrt{10}}$, $\theta = 63.6°$ (1 d.p.)

5. $u^2 = 2x - 1$, $x = \tfrac{1}{2}(u^2 + 1)$, $2u \dfrac{du}{dx} = 2$, so $dx = u\,du$

$I = \int \tfrac{1}{2}(u^2 + 1)(\sqrt{u^2})u\,du = \int \tfrac{1}{2}(u^4 + u^2)\,du$

$= \tfrac{1}{10}u^5 + \tfrac{1}{6}u^3 + c$

$= \tfrac{1}{30}u^3(3u^2 + 5) + c$

$= \tfrac{1}{30}(2x - 1)^{\frac{3}{2}}(6x + 2) + c$

$= \tfrac{1}{15}(2x - 1)^{\frac{3}{2}}(3x + 1) + c$

C3 Practice paper

1. $\dfrac{4}{x - 1} - \dfrac{2x - 6}{2x^2 - 5x + 3}$

$= \dfrac{4}{x - 1} - \dfrac{2x - 6}{(2x - 3)(x - 1)}$

$= \dfrac{8x - 12 - 2x + 6}{(2x - 3)(x - 1)}$

$= \dfrac{6x - 6}{(2x - 3)(x - 1)}$

$= \dfrac{6(x - 1)}{(2x - 3)(x - 1)}$

$= \dfrac{6}{2x - 3}$

2. (a) $18\,\text{mm}^2$

(b) 12.26 pm (accept 26 minutes)

3. (a)

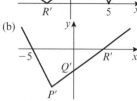

(b)

(c) $f(x) = |x - 2| - 3$

Minimum value of $f(x)$ is -3 when $x = 2$, so P is $(2, -3)$

When $x = 0$, $f(x) = |-2| - 3 = 2 - 3 = -1$,

so Q is $(0, -1)$

When $f(x) = 0$, $|x - 2| = 3$, so $x = 5$ or -1, so R is $(-1\ 0)$

4. (i) $\dfrac{dy}{dx} = \tfrac{3}{2}e^{3x}(2x^2 + 1)^{\frac{1}{2}}4x + (2x^2 + 1)^{\frac{3}{2}}3e^{3x}$

When $x = 0$, $\dfrac{dy}{dx} = 3$

(ii) $\dfrac{dy}{dx} = \dfrac{3 + 8x - 3x^2}{(x^2 + 1)^2}$

Turning points when $\dfrac{dy}{dx} = 0$, at $x = -\tfrac{1}{3}$ and $x = 3$

5. (a) $f(3) = 3\ln 3 - 2\sqrt{3} = -0.168\,26...$

$f(4) = 3\ln 4 - 4 = 0.158\,88...$

Sign change, so root between 3 and 4

(b) $3\ln x = 2\sqrt{x}$, $\ln x = \tfrac{2}{3}\sqrt{x}$, $e^{\frac{2}{3}\sqrt{x}} = x$

(c) $x_1 = 3.1731$, $x_2 = 3.2790$, $x_3 = 3.3441$

(d) $f(3.4485) = -0.000\,209\,402$

$f(3.4495) = 0.000\,121\,956$

Sign change, so $3.4485 < \alpha < 3.4495$, i.e. $\alpha = 3.449$ (3 d.p.)

6. (a) LHS $= x = \sqrt{3}\pi$

RHS $= 6 \times \dfrac{\pi}{6} \times \tan\dfrac{\pi}{3} = \pi \times \sqrt{3}$ so P lies on C

(b) $\dfrac{dx}{dy} = 12y\sec^2 2y + 6\tan 2y$

When $y = \dfrac{\pi}{6}$, $\dfrac{dx}{dy} = 8\pi + 6\sqrt{3}$

Equation of tangent at P is $(8\pi + 6\sqrt{3})y = x + \tfrac{4}{3}\pi^2$

7. (a) $f(x) = x^2 - 4x + 3 = (x - 2)^2 - 4 + 3 = (x - 2)^2 - 1$

Since $(x - 2)^2 \geqslant 0$, the range of $f(x)$ is $f(x) \geqslant -1$

(b) Let $y = f(x)$ i.e. $y = (x - 2)^2 - 1$ so $y + 1 = (x - 2)^2$

$x = 2 \pm \sqrt{(y + 1)}$ ($\pm$ sign must be included at this stage)

So $f^{-1}(x) = 2 + \sqrt{(x + 1)}$ (+ sign only now because

range of $f^{-1}(x) = $ domain of $f(x)$ i.e. $x \geqslant 2$)

(c) $gf(x) = |x^2 - 4x + 3 - 4|$ or $|x^2 - 4x - 1|$

(d) $x^2 - 4x - 1 = 4$ giving $x = 5$ and $x = -1$ (inadmissible)

or $x^2 - 4x - 1 = -4$ giving $x = 3$ and $x = 1$ (inadmissible)

So solutions are $x = 3$ and $x = 5$

8. (a) $11\sin\theta + 7\cos\theta = R\cos\theta\cos\alpha + R\sin\theta\sin\alpha$

$R\sin\alpha = 11$, $R\cos\alpha = 7$, $\tan\alpha = \tfrac{11}{7}$, $\alpha = 57.5°$

$R = \sqrt{11^2 + 7^2} = \sqrt{170}$

So $11\sin\theta + 7\cos\theta = \sqrt{170}\cos(\theta - 57.5°)$

(b) $\sqrt{170}\cos(\theta - 57.5°) = 12$

$\cos(\theta - 57.5°) = \dfrac{12}{\sqrt{170}}$

$-57.5° \leqslant (\theta - 57.5°) \leqslant 302.5°$

$\theta - 57.5° = -23.0°$ or $23.0°$

So $\theta = 34.5°$ or $80.5°$

9. (a) $(\sin 67.5° + \cos 67.5°)^2$

$= \sin^2 67.5° + \cos^2 67.5° + 2\sin 67.5°\cos 67.5°$

$= 1 + \sin 135°$

$= 1 + \dfrac{1}{\sqrt{2}}$

(b) $2\cos 4x = 1 - 4\cos 2x$

$2(2\cos^2 2x - 1) = 1 - 4\cos 2x$

$4\cos^2 2x + 4\cos 2x - 3 = 0$

$(2\cos 2x + 3)(2\cos 2x - 1) = 0$

$\cos 2x = -\tfrac{3}{2}$ or $\cos 2x = \tfrac{1}{2}$

Range for $2x$ is $0 \leqslant 2x \leqslant 4\pi$

So $2x = \dfrac{\pi}{3}, \dfrac{5\pi}{3}, \dfrac{7\pi}{3}, \dfrac{11\pi}{3}$

i.e. $x = \dfrac{\pi}{6}, \dfrac{5\pi}{6}, \dfrac{7\pi}{6}, \dfrac{11\pi}{6}$

C4 Practice paper

1. (a)

x	0	$\dfrac{\pi}{4}$	$\dfrac{\pi}{2}$	$\dfrac{3\pi}{4}$	π
y	4	3.695 52	2.828 43	1.530 73	0

(b) Estimate of area of R

$= \dfrac{\pi}{8}[4 + 0 + 2(3.695\,52 + 2.828\,43 + 1.530\,73)]$

$= \dfrac{\pi}{8}(20.109\,36)$

$= 7.8970$ (4 d.p.)

(c) $\displaystyle\int_0^\pi 4\cos\left(\dfrac{x}{2}\right)dx = \left[8\sin\left(\dfrac{x}{2}\right)\right]_0^\pi = 8 - 0 = 8$

(d) % error $= \dfrac{8 - 7.8969}{8} \times 100 = 1.288\,75\%$

2. $u = 4 - x^3$, $\dfrac{du}{dx} = -3x^2$, $-\dfrac{1}{3}du = x^2\,dx$, $x^3 = 4 - u$

When $x = 0$, $u = 4$, and when $x = 1$, $u = 3$

$\displaystyle\int_0^1 \dfrac{x^5}{4 - x^3}dx = \int_0^1 \dfrac{x^3 x^2}{4 - x^3}dx = \int_4^3 \dfrac{(4 - u)}{u}\left(-\dfrac{1}{3}\right)du = \dfrac{1}{3}\int_3^4\left(\dfrac{4}{u} - 1\right)du$

$= \dfrac{1}{3}\Big[4\ln u - u\Big]_3^4 = \dfrac{1}{3}[(4\ln 4 - 4) - (4\ln 3 - 3)]$

$= \dfrac{1}{3}\left[4\ln\dfrac{4}{3} - 1\right] = \dfrac{4}{3}\ln\left(\dfrac{4}{3}\right) - \dfrac{1}{3}$

3. (a) $\dfrac{8x + 2}{(1 - 3x)(2 + x)} = \dfrac{A}{1 - 3x} + \dfrac{B}{2 + x}$

$8x + 2 = A(2 + x) + B(1 - 3x)$

When $x = -2$, $-14 = 7B$, so $B = -2$

When $x = \dfrac{1}{3}$, $\dfrac{14}{3} = \dfrac{7}{3}A$, so $A = 2$

So $\dfrac{8x + 2}{(1 - 3x)(2 + x)} = \dfrac{2}{1 - 3x} - \dfrac{2}{2 + x}$

(b) $f(x) = 2(1 - 3x)^{-1} - 2\left[2^{-1}\left(1 + \dfrac{x}{2}\right)^{-1}\right]$

$= 2(1 - 3x)^{-1} - \left(1 + \dfrac{x}{2}\right)^{-1}$

$= 2(1 + 3x + 9x^2...) - \left(1 - \dfrac{x}{2} + \dfrac{x^2}{4}...\right)$

$= 1 + \dfrac{13x}{2} + \dfrac{71x^2}{4}...$

4. (a) Let $u = t$ and $\dfrac{dv}{dt} = e^t$, then $\dfrac{du}{dt} = 1$ and $v = e^t$

So $\int t\,e^t\,dt = t\,e^t - \int e^t \times 1\,dt = t\,e^t - e^t + c$

(b) Let $I = \int(t - t^2)\,e^t\,dt = \int t\,e^t\,dt - \int t^2\,e^t\,dt$

$\int t^2\,e^t\,dt = t^2\,e^t - \int e^t\,2t\,dt = t^2\,e^t - 2(t\,e^t - e^t)$

So $I = (t\,e^t - e^t) - [t^2\,e^t - 2(t\,e^t - e^t)] = 3t\,e^t - t^2\,e^t - 3e^t$

and so (putting in the limits)

$\displaystyle\int_0^1(t - t^2)\,e^t\,dt = \Big[3t\,e^t - t^2\,e^t - 3e^t\Big]_0^1 = 3 - e$

5. (a) Area of $R = \displaystyle\int_{-4}^0 \dfrac{5}{\sqrt{1 - 2x}}dx = 5\int_{-4}^0 (1 - 2x)^{-\frac{1}{2}}dx$

$= 5\Big[-(1 - 2x)^{\frac{1}{2}}\Big]_{-4}^0 = 5[(-1) - (-3)] = 10$

(b) Volume $= \displaystyle\int \pi y^2\,dx = \pi\int_{-4}^0 \dfrac{25}{(1 - 2x)}dx$

$= 25\pi\Big[-\tfrac{1}{2}\ln(1 - 2x)\Big]_{-4}^0 = 25\pi\Big[0 - (-\tfrac{1}{2}\ln 9)\Big]$

$= \dfrac{25\pi}{2}\ln 9 = 25\pi \ln 3$

6. (a) $3y\,e^{-x} = 4x + y^2$

Differentiating implicitly:

$(3y)(-e^{-x}) + (e^{-x})\left(3\dfrac{dy}{dx}\right) = 4 + 2y\dfrac{dy}{dx}$

$(3e^{-x} - 2y)\dfrac{dy}{dx} = 4 + 3y\,e^{-x}$

$\dfrac{dy}{dx} = \dfrac{4 + 3y\,e^{-x}}{3e^{-x} - 2y}$

(b) $P(0, 3)$ lies on the curve

$\dfrac{dy}{dx}$ at $P = \dfrac{4 + 9e^0}{3e^0 - 6} = \dfrac{4 + 9}{3 - 6} = -\dfrac{13}{3}$

Gradient of normal at $P = \dfrac{3}{13}$

Equation of normal at P is $y - 3 = \dfrac{3}{13}(x - 0)$

$13y - 39 = 3x$

$3x - 13y + 39 = 0$

7. (a) $\dfrac{dx}{dt} = \dfrac{-kt^2 e^{\frac{x}{3}}}{2}$ i.e. $\int e^{-\frac{x}{3}}dx = -\dfrac{k}{2}\int t^2\,dt$

which integrates to $-3\,e^{-\frac{x}{3}} = \dfrac{-kt^3}{6} + c$

When $t = 0$, $x = 6$, which gives $c = -3e^{-2}$

So $-3\,e^{-\frac{x}{3}} = \dfrac{-kt^3}{6} - 3e^{-2}$ i.e. $e^{-\frac{x}{3}} = \dfrac{kt^3}{18} + e^{-2}$

i.e. $-\dfrac{x}{3} = \ln\left(\dfrac{kt^3}{18} + e^{-2}\right)$ or $x = -3\ln\left(\dfrac{kt^3}{18} + e^{-2}\right)$

(b) $k = 0.009$ so $x = -3\ln(0.0005t^3 + e^{-2})$

(i) When $t = 8$, $x = -3\ln(0.0005 \times 512 + e^{-2})$

$= 2814.57...$

so population $= 2800$ (to nearest 100)

(ii) When $x = 0$, $-3\ln(0.0005t^3 + e^{-2}) = 0$

i.e. $0.0005t^3 + e^{-2} = 1$

$t^3 = \dfrac{(1 - e^{-2})}{0.0005} = 1729.329...$ i.e. $t = 12.0030...$

$= 12$ minutes (to nearest minute)

8. (a) $7 + 5\lambda = -6 + 4\mu$

$6 - \lambda = 1 + 3\mu$

$-5 - 3\lambda = 2 - 2\mu$

Solving gives $\lambda = -1$ and $\mu = 2$

Position vector of P is $\begin{pmatrix} 2 \\ 7 \\ -2 \end{pmatrix}$

(b) Let θ be the angle between the direction vectors

$\begin{pmatrix} 5 \\ -1 \\ -3 \end{pmatrix}$ and $\begin{pmatrix} 4 \\ 3 \\ -2 \end{pmatrix}$

Magnitudes are $\sqrt{35}$ and $\sqrt{29}$ respectively, scalar product $= 23$

$\cos\theta = \dfrac{23}{\sqrt{35}\sqrt{29}}$, giving $\theta = 43.8°$

(c) Position vector of A is $\begin{pmatrix} 12 \\ 5 \\ -8 \end{pmatrix}$

(d) $\overrightarrow{AP} = \overrightarrow{AO} + \overrightarrow{OP} = \begin{pmatrix} -12 \\ -5 \\ 8 \end{pmatrix} + \begin{pmatrix} 2 \\ 7 \\ -2 \end{pmatrix} = \begin{pmatrix} -10 \\ 2 \\ 6 \end{pmatrix}$

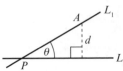

Distance $AP = |\overrightarrow{AP}| = \sqrt{140}$

$d = |\overrightarrow{AP}|\sin\theta = \sqrt{140} \times \sin 43.8° = 8.19$ (3 s.f.)

For your own notes

. . . .

. . . .

. . . .

. . . .

. . . .

. . . .

. . . .

. . . .

. . . .

. . . .

. . . .

. . . .

. . . .

. . . .

. . . .

. . . .

. . . .

. . . .

. . . .

. . . .

. . . .

. . . .

. . . .

. . . .

. . . .

. . . .

. . . .

. . . .

. . . .

. . . .

For your own notes

. . . .

. . . .

Published by Pearson Education Limited, 80 Strand, London, WC2R 0RL.

www.pearsonschoolsandfecolleges.co.uk

Copies of official specifications for all Edexcel qualifications may be found on the website: www.edexcel.com

Text and illustrations © Pearson Education Limited 2016

Edited by Project One Publishing Solutions, Scotland

Typeset and illustrations by Tech-Set Ltd, Gateshead

Cover illustration by Miriam Sturdee

The right of Glyn Payne to be identified as author of this work has been asserted by him in accordance with the Copyright, Designs and Patents Act 1988.

First published 2016

18 17 16

10 9 8 7 6 5 4 3 2 1

British Library Cataloguing in Publication Data
A catalogue record for this book is available from the British Library

ISBN 9781292111162

Printed in Slovakia by Neografia